Cambridge Elements

Elements in the Structure and Dynamics of Complex Networks
edited by
Guido Caldarelli
Ca' Foscari University of Venice

MODULARITY AND DYNAMICS ON COMPLEX NETWORKS

Renaud Lambiotte
University of Oxford

Michael T. Schaub
RWTH Aachen University

CAMBRIDGE
UNIVERSITY PRESS

University Printing House, Cambridge CB2 8BS, United Kingdom

One Liberty Plaza, 20th Floor, New York, NY 10006, USA

477 Williamstown Road, Port Melbourne, VIC 3207, Australia

314–321, 3rd Floor, Plot 3, Splendor Forum, Jasola District Centre,
New Delhi – 110025, India

103 Penang Road, #05–06/07, Visioncrest Commercial, Singapore 238467

Cambridge University Press is part of the University of Cambridge.

It furthers the University's mission by disseminating knowledge in the pursuit of
education, learning, and research at the highest international levels of excellence.

www.cambridge.org
Information on this title: www.cambridge.org/9781108733533
DOI: 10.1017/9781108774116

First published 2021

A catalogue record for this publication is available from the British Library.

ISBN 978-1-108-73353-3 Paperback
ISSN 2516-5763 (online)
ISSN 2516-5755 (print)

Modularity and Dynamics on Complex Networks

Elements in the Structure and Dynamics of Complex Networks

DOI: 10.1017/9781108774116
First published online: December 2021

Renaud Lambiotte
University of Oxford

Michael T. Schaub
RWTH Aachen University

Authors for correspondence: Renaud Lambiotte,
renaud.lambiotte@some.ox.ac.uk;
Michael T. Schaub, schaub@cs.rwth-aachen.de

Abstract: Complex networks are typically not homogeneous, as they tend to display an array of structures at different scales. A feature that has attracted a lot of research is their modular organisation (i.e., networks may often be considered as being composed of certain building blocks, or modules). In this Element, the authors discuss a number of ways in which this idea of modularity can be conceptualised, focusing specifically on the interplay between modular network structure and dynamics taking place on a network. They discuss, in particular, how modular structure and symmetries may impact on network dynamics and, vice versa, how observations of such dynamics may be used to infer the modular structure. They also revisit several other notions of modularity that have been proposed for complex networks and show how these can be related to and interpreted from the point of view of dynamical processes on networks.

Keywords: modularity, networks, time scale, dynamics, block models

ISBNs: 9781108733533 (PB), 9781108774116 (OC)
ISSNs: 2516-5763 (online), 2516-5755 (print)

Contents

1 Introduction

Over the last 20 years, networks and graphs have become a near-ubiquitous modelling framework for complex systems. By representing the entities of a system as nodes in a graph and encoding relationships between these nodes as edges, we can abstract systems from a variety of domains with the same mathematical language, including biological, social, and technical systems (Newman, 2018a). Network abstractions are often used with one of the two following perspectives in mind. First, graphs and networks provide a natural way to describe relational data (i.e., datasets corresponding to 'interactions' or correlations between pairs of entities). A prototypical example here is online social networks, in which we measure interactions between actors and can derive a network representation of the social system based on these measurements. We may then try to explain certain properties of the social system by modelling and analysing the network (e.g., by searching for interesting connectivity patterns between the nodes). Second, networks are often used to describe distributed dynamical systems, including prototypical examples such as power grids, traffic networks, or various other kinds of supply or distribution networks. The edges of the network are in this context not the primary object of our modelling. Rather, we are interested in understanding a dynamical process that takes place on this network. More specifically, we often aim to comprehend how the network structure shapes this dynamics (e.g., in terms of its long-term behaviour). In reality, of course, both these perspectives are simplifications in that for many real systems, there are typically uncertainty and dynamics associated to both node and edge variables which make up the network: think, for instance, of a rumour spreading on a social network, where both node variables (the infection state) and the network edges (who is in contact with whom) will be highly dynamic and uncertain. We may not know the exact status of each individual; moreover, edges will change dynamically, and their presence or absence may not be determined accurately.

No matter under what perspective we are interested in networks, it should be intuitively clear that networks with some kind of 'modular structure' may be of interest to us. For now, consider modular structure simply in terms of a network made of dense clusters that are loosely connected with each other. From the perspective of relational data, modular structure may be indicative of a hidden cause that binds a set of nodes together: this corresponds to the idea of homophily in social networks (McPherson, Smith-Lovin, & Cook, 2001), which can lead to the formation of communities of tightly knit actors. From the perspective of dynamics, it is often impractical to keep a full description of a dynamical process on a network when the number of dynamical units is too large. In many cases, it is unclear whether such finely detailed data is necessary to understand the global phenomena of interest, as relevant observables can often be obtained by aggregating microscopic information into macroscopic information (i.e., aggregating information over many nodes). This kind of dimensionality reduction of the dynamics is particularly successful if there exist roughly homogeneously connected blocks of nodes (i.e., a modular network structure (Simon & Ando, 1961)).

As the title indicates, this Element will primarily adopt a dynamical perspective on network analysis. Accordingly, our core objective will be to explore the relations between modular structure and dynamics on networks; but we will also explain how certain aspects of the analysis of relational data can be interpreted from this lens. However, an exhaustive exposition of methods to characterise and uncover modules (also called blocks, clusters, or communities) in networks will not be the main focus of our exposition. We refer the interested reader to the extensive literature on this topic for more detailed treatments; see, for example, Doreian, Batagelj, and Ferligoj (2020); Fortunato and Hric (2016); Newman (2018a); Schaeffer (2007).

Network Dynamics and Network Structure

It is well-known that there exists a two-way relationship between dynamics on graphs and the underlying graph structure (Schaub et al., 2019b). On the one hand, the structure of a network affects dynamical processes taking place on it (Porter & Gleeson, 2016). In the simplest case of a linear dynamical system, this relationship derives from the spectral properties of a matrix encoding the graph, most often the adjacency matrix or the graph Laplacian. On the other hand, dynamics can help reveal salient features of a graph. This includes the identification of central nodes or the detection of modules in large-scale networks.

To illustrate this two-fold relation between network structure and network dynamics on an intuitive level, let us consider random walks on networks. Random walks are often used as a model for diffusion, and there is much research on the impact of network structure on different properties of random-walk dynamics (Masuda, Porter, & Lambiotte, 2017). In particular, degree heterogeneity, finite size effects and modular structure can all make diffusion processes on networks quantitatively and even qualitatively different from diffusion on regular or infinite lattices. At the same time, random walks are key to many algorithms that uncover various types of structural properties of networks. For example, the classical PageRank method for identifying important nodes may be interpreted in terms of a random walk(Gleich, 2015). Indeed, as we will discuss, several algorithms use trajectories of dynamical processes such as random walks to reveal mesoscale network patterns.

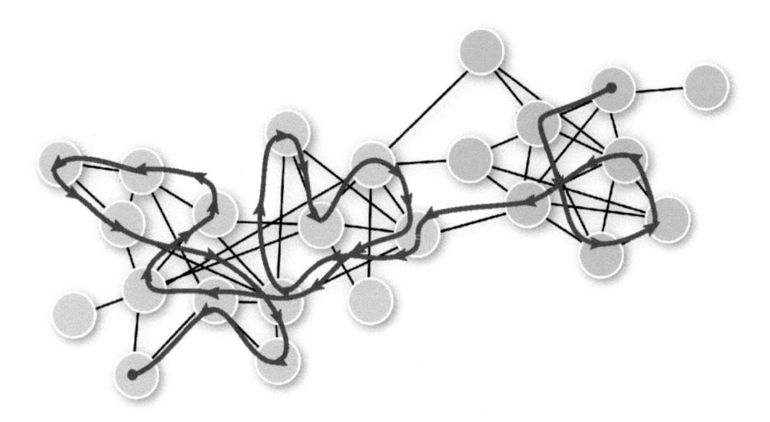

Figure 1 Modularity and Dynamics on Networks. Our main ambition is to understand relationships between modular structure of a network, here highlighted in different node colours, and a dynamics taking place on it, here illustrated with the red trajectory on the network. The two complementary questions at the core of this Element are: (1) How does the modular structure of a network affect dynamics? (2) How can dynamics help us characterise and uncover the modular structure of a network?

Outline of This Element

In this Element, we try to provide a basic overview of the topic of modularity and dynamics on complex networks. Our exposition is structured as follows. In Section 2, we first discuss some background material in Network Science and then review classical notions of modular structure in networks in Section 3. In Sections 4 and 5, we discuss the interplay between dynamics and network structure in terms of timescale separation and symmetries, and how these aspects can be used to reduce the complexity of the description of network dynamics. In Section 6, we then explain how we can detect so-called assortative community structure, primarily based on the notion of timescale separation. Section 7 then discusses the definition and detection of more general (dynamical) block structure, leveraging ideas from linear systems theory and symmetry reduction. In Section 8, we conclude with a short discussion on avenues for future work and additional perspectives.

Why Are Networks Modular?

For many years, researchers have been fascinated by the prevalence of modularity in systems as different as the World Wide Web, foodwebs, and brain networks, raising the question: are there universal mechanisms driving the evolution of networks toward a modular architecture? Among the many mechanisms that have been proposed (Meunier, Lambiotte, & Bullmore, 2010), the following profound idea of Herbert Simon (1962) stands out by its elegance. 'Nearly-decomposable' systems, as Simon calls them, allow faster adaptation or evolution of the system in response to changing environmental conditions. In Simon's view, modules represent stable building blocks that ensure the robustness of a system evolving under changing or competitive selection criteria. To illustrate this idea, Simon wrote an intuitive parable about two watchmakers, called 'Hora' and 'Tempus' (Simon, 1962):

> There once were two watchmakers, named Hora and Tempus, who manufactured very fine watches. Both of them were highly regarded, and the phones in their workshops rang frequently – new customers were constantly calling them. However, Hora prospered, while Tempus became poorer and poorer and finally lost his shop. What was the reason?
>
> The watches the men made consisted of about 1,000 parts each. Tempus had so constructed his that if he had one partly assembled and had to put it down – to answer the phone say – it immediately fell to pieces and had to be reassembled from the elements. The better the customers liked his watches, the more they phoned him, the more difficult it became for him to find enough uninterrupted time to finish a watch.
>
> The watches that Hora made were no less complex than those of Tempus. But he had designed them so that he could put together subassemblies of about ten elements each. Ten of these subassemblies, again, could be put together into a larger subassembly; and a system of ten of the latter subassemblies constituted the whole watch. Hence, when Hora had to put down a partly assembled watch in order to answer the phone, he lost only a small part of his work, and he assembled his watches in only a fraction of the man-hours it took Tempus.

This story illustrates in simple terms the potential evolutionary advantage that a modular system structure may have, and provides an argument for the ubiquity of modularity in a broad range of natural and social systems.[a] In the following, we will not dwell on why there is modular structure in the network, but rather focus on the question: how does the modularity of a network affect its behaviour and, in particular, its dynamical properties?

[a] One needs to be careful with such statements. Simon himself cautioned that many systems lack conclusive statistical evidence for being modular and may only be perceived as modular due to confirmation bias. However, the statement that many networks are modular has been validated on a large corpus of network datasets by now. See, for example, Fortunato (2010); Ghasemian, Hosseinmardi, and Clauset (2019); Leskovec et al. (2008).

Table 1 Notation

Symbol	Description
$n \in \mathbb{N}$	number of nodes
$m \in \mathbb{R}$	total weight of edges (number of edges for unweighted networks)
$\mathcal{C} \in \mathbb{N}$	number of modules / communities
$\mathcal{V} = \{1, \ldots, n\}$	set of nodes / vertices
$i, j, \ell \in \{1, \ldots, n\}$	indices for nodes
$\mathcal{P} = \{\mathcal{A}_1, \ldots, \mathcal{A}_{\mathcal{C}}\}$	partition of the nodes into communities $\mathcal{A}_\alpha$
$\mathcal{A}_\alpha$	set of nodes within the αth community
$\alpha, \beta \in \{1, \ldots, \mathcal{C}\}$	indices for communities
$k_i \in \mathbb{R}$	weighted degree (strength) of node i
$\boldsymbol{A} \in \mathbb{R}^{n \times n}$	weighted adjacency matrix of a network
$\boldsymbol{K}(\boldsymbol{A}) = \mathrm{diag}(\boldsymbol{A}\mathbf{1}) \in \mathbb{R}^{n \times n}$	weighted degree matrix of a network
$\boldsymbol{L}(\boldsymbol{A}) = \boldsymbol{K} - \boldsymbol{A}$	combinatorial Laplacian matrix
$\boldsymbol{\mathcal{L}}(\boldsymbol{A}) = \boldsymbol{I} - \boldsymbol{K}^{-1/2}\boldsymbol{A}\boldsymbol{K}^{-1/2}$	normalised Laplacian matrix
$\boldsymbol{L}_{\mathrm{rw}}(\boldsymbol{A}) = \boldsymbol{I} - \boldsymbol{K}^{-1}\boldsymbol{A}$	random-walk Laplacian matrix
$\boldsymbol{H} \in \{0, 1\}^{n \times \mathcal{C}}$	partition indicator matrix with entries $H_{i\alpha} = 1$, if node i is in the αth community ($\mathcal{A}_\alpha$), and $H_{i\alpha} = 0$ otherwise
$\boldsymbol{h}_\alpha \in \{0, 1\}^n$	Indicator vector of the αth community (i.e., αth column of $\boldsymbol{H}$)
$\gamma: \{1, \ldots, n\} \to \{1, \ldots, n\}$	permutation function of node labels
$\boldsymbol{\Gamma}$	permutation matrix associated to γ
$d(x, y)$	distance function between x and y
$\kappa(x, y)$	kernel function of x and y

Notation

We use the following general mathematical notations and conventions. We denote vectors by small letters in bold such as $\boldsymbol{x}, \boldsymbol{y}$ and use $(\cdot)^\top$ to denote the transpose of a vector. Our convention is that all vectors are column vectors, and accordingly, $\boldsymbol{x}^\top$ is a row vector. We use $\mathbf{1}$ to denote the vector of all ones. Matrices are denoted by bold uppercase letters such as $\boldsymbol{A}, \boldsymbol{M}$, where $\boldsymbol{I}$ is used to denote the identity matrix. We write $\mathrm{diag}(\boldsymbol{x})$ to denote the diagonal matrix whose diagonal entries are defined by the components of vector $\boldsymbol{x}$ and are 0 otherwise. Entries of vectors or matrices are non-bold with subscripts. For instance, vector $\boldsymbol{x}$ has entries $x_1, \ldots, x_n$ and the matrix $\boldsymbol{A}$ has entries A_{ij}. If there is ambiguity, we may alternatively use the notation $[\boldsymbol{v}_i]_j$ to denote the jth entry of a vector $\boldsymbol{v}_i$ (or a matrix, accordingly). Finally, we use $\mathbb{P}(\cdot)$ and $\mathbb{E}[\cdot]$ for the probability and expectation of the statement inside the parentheses, respectively.

More specific notation regarding networks and associated objects is summarised in Table 1. These objects are explained in Section 2.

2 Background Material

In this section, we review some notions from algebraic and spectral graph theory as well as the theory of linear dynamical systems. These concepts will be essential for our discussions in the following sections.

2.1 Graph Theory

Networks provide a natural framework to represent systems composed of elements in interaction. At the core of a network representation is the inherent assumption that the system under investigation can be decomposed into nodes, representing the system elements, and edges, representing pairwise interactions between the system elements.

In the simplest setting, we assume that both the nodes and the edges of a network are all of the same type and their number is fixed. All of these assumptions can be relaxed,[1] but we will be mostly concerned with undirected (and weighted) networks in this Element. Within this setup we can represent a network mathematically by a graph $\mathcal{G}(\mathcal{V}, \mathcal{E})$, with a set of nodes $\mathcal{V}$ of cardinality $n := |\mathcal{V}|$, and a set of edges $\mathcal{E} = \{\{i, j\} \mid i, j \in \mathcal{V}\}$. Without loss of generality, we will identify the node set $\mathcal{V}$ with the set $\{1, \ldots, n\}$. For a weighted graph, we endow the graph $\mathcal{G}$ with a weight function $w_{\mathcal{G}}: \mathcal{E} \to \mathbb{R}_+$, which maps each edge $\{i, j\}$ to a positive weight, $w_{\mathcal{G}}(\{i, j\}) = w_{ij}$.

More generally, we can consider directed graphs, meaning that node i may be adjacent (connected) to node j but not vice versa. This lack of symmetry leads to a number of mathematical complications that make directed networks and dynamical systems acting on directed networks far more difficult to analyse (cf. the box 'The Case of Directed Networks' in Section 2.3.3). We will thus focus on undirected networks, unless otherwise stated.

The edge set of a graph describes which nodes are adjacent (i.e., directly connected by an edge). Especially in the context of dynamical systems defined on graphs, we need to capture how a sequence of direct connections defines indirect connectivity between pairs of nodes, leading to the additional notions

[1] For simplicity, within this Element, we will concentrate on undirected graphs with positive edges weights and provide some additional discussion on directed networks. Relaxing the above modelling assumptions leads to various notions, including signed networks (Kunegis et al., 2010), multiplex networks (Kivela et al., 2014), temporal networks (Holme & Saramäki, 2019), and higher-order networks (Battiston et al., 2020; Lambiotte, Rosvall, & Scholtes, 2019; Schaub et al., 2021). These are active areas of research that we will mention further when relevant, and we invite the reader to consult the literature for further information on these topics.

of a walk, trail, path, and connectedness of a graph. A walk between a starting node i and a terminal node j is a sequence of edges such that there exists an associated node sequence $(i, \ell, \dots, j)$, in which every subsequent pair of nodes is adjacent. A trail is a walk in which all edges are distinct, and a path is a trail in which additionally all nodes are distinct. A graph is connected if there is a path between any two nodes. When a graph is not connected, it is composed of several connected components.[2]

There are different ways to describe a graph algebraically. One representation that will attract a lot of our attention is the so-called adjacency matrix. The adjacency matrix $\boldsymbol{A}$ is an $n \times n$ matrix encoding the presence or absence of an edge between any pair of nodes in the graph. Its entries are

$$A_{ij} = \begin{cases} 1 & \text{if node } i \text{ is adjacent to node } j, \\ 0 & \text{otherwise.} \end{cases} \tag{2.1}$$

If the network is weighted, then $A_{ij} = w_{ij}$ if there is an edge between i and j and zero otherwise. Here w_{ij} is the edge weight associated to edge $\{i, j\}$, as discussed earlier. Clearly, for undirected graphs we have $A_{ij} = A_{ji}$ (i.e., the adjacency matrix is symmetric ($\boldsymbol{A} = \boldsymbol{A}^\top$)). Using the adjacency matrix, we can express the weighted degree of each node i as $k_i = \sum_j A_{ij}$. The degree of a node is equal to the number of neighbours of a node for simple, unweighted graphs.

2.2 Random Graph Models

Many local or global properties of a network can potentially be of interest, whether in terms of their influence on a dynamics acting on a network or otherwise. For instance, let us consider the clustering coefficient C defined as

$$C := \frac{\#\text{ triangles in graph}}{(\#\text{ number of connected triplets})/3}, \tag{2.2}$$

which measures the relative abundance of triangles in a network. By construction, the value of the clustering coefficient C lies between 0 and 1, and it is often considered as a measure of the cohesion inside a network.

Now suppose that we observe a value of the clustering coefficient of $C = 0.25$ in an empirical network. Should we conclude that this value is small or large? In order to answer this question and interpret our measurement in a meaningful way, we often require some suitable reference points for the measurement. Such reference points are often deduced from random graph models.

[2] Note that a single node without any connection is a trivial connected component by definition.

A random graph model defines an ensemble of graphs (i.e., a set of possible graphs and a probability associated to each of those graphs). Random graph models are often defined by considering the edges in the graph as random variables, which obey certain probabilistic laws. One of the simplest random graph models is the Erdős–Rényi (ER) model, also called the Poisson or binomial random graph. The Erdős–Rényi random graph has two parameters: the number of nodes n, and the probability q that a link exists between a pair of nodes. By definition, self-loops are excluded. Each pair of nodes (undirected edge) is then seen as an independent Bernoulli random variable that determines the presence or absence of a link. The combined realisations of all these $n(n-1)/2$ Bernouilli random variables then determine one realisation of the random graph. Note that any network without multiple edges and self-loops can be generated by sampling from an ER model as long as $0 < q < 1$. However, different realisations will be observed with different probabilities, depending on the value of q. Exploiting the fact that each link exists independently with the same probability, several properties of the ER model can be calculated analytically, in particular the expected value of several network metrics as well as their variance.

However, thc ER model is known to produce unrealistic edge patterns. In particular, it generates networks with a Poisson degree distribution, which is not typical for most observed networks in the real world.[3] For this reason, other random graph models such as the configuration model are often considered to be a more appropriate baseline for real-world networks. The configuration model (Fosdick et al., 2018) is defined as a random graph model in which all possible configurations appear with the same probability under the constraint that each node i has a given degree k_i ($1 \leq i \leq n$). Hence, the configuration model generates an ensemble of random graphs with a prescribed degree distribution that can either be taken from empirical data or chosen from a family of functions (e.g., a power-law distribution). The soft configuration model, or Chung-Lu model (F. F.Chung & Lu, 2002; Park & Newman, 2004), is defined analogously to the configuration model; but rather than fixing the exact degree sequence of the graph, only the expected degree sequence is prescribed. Similar to the ER model, several properties of the (soft) configuration model can be calculated analytically. Of interest for the following sections is the expected number of links between two nodes i and j in the soft-configuration model, which for $k_i k_j < 2m$ is given by

[3] Power-law or not (Broido & Clauset, 2019), many real-world networks tend to exhibit a degree distribution with a fat tail: the vast majority of nodes have only a small degree, and a small number of hubs have a large number of connections.

$$\mathbb{E}[A_{ij}] = \frac{k_i k_j}{2m}. \tag{2.3}$$

Here $\mathbb{E}[A_{ij}]$ represents the expected value of the adjacency matrix for the edge $\{i, j\}$, which is simply the probability $\mathbb{P}(A_{ij} = 1)$ for an unweighted graph. As can be seen, this probability for an edge to exist between a pair of nodes is clearly not uniform, which was the case in the ER model. Note that here and in the following we are interested in the original soft configuration model proposed by Chung and Lu, as specified earlier, though a more general treatment can be given for graphs where the product $k_i k_j$ can exceed $2m$ (Park & Newman, 2004).

Let us now return to the initial motivating example of this section (i.e., determining if a specific statistic like the clustering coefficient is small or large in a real-world network). A common practice is to consider how this measure would be statistically distributed for graphs drawn from a soft configuration model with an expected degree sequence matching the empirically analysed network. We can then determine if the empirical value is significantly different from that of the random graph model, for instance by calculating a Z score or a p value. For the specific case of the clustering coefficient, its expected value under the configuration model is given by M. E. J. Newman (2018a):

$$C = \frac{\left(\mathbb{E}[k^2] - \mathbb{E}[k]\right)^2}{\mathbb{E}[k]^3 n}. \tag{2.4}$$

Note that this value is extremely small for large values of n unless the variance of the degree diverges. Thus, we would expect a very small clustering coefficient for large networks. This type of approach is popular for motif analysis, where the purpose is to uncover important motifs in a network, whose over- or under-representation may be associated to their function in the system (Milo et al., 2002). Let us emphasise here that this conclusion strongly depends on the choice of model, as the question of whether a particular statistic is significant can only be answered with respect to the chosen random model. Other models may have very different behaviour, and we thus need to exercise caution when declaring that some network property is significant or not.

2.3 Network Dynamical Systems and Linear Dynamics

2.3.1 Linear Dynamics on Networks

In many situations, the nodes of a network are not static entities, but each node i carries a state $x_i(t)$ that evolves in either continuous or discrete time. Consider, for instance, the formation of opinions in a social network, where each node i may update its opinion $x_i(t)$ on a particular topic based on the interactions with adjacent nodes or, alternatively, on the dynamics of a set of connected neurons in a part of the brain.

While each node has a dynamical state, it is typically assumed that the network is static when considering such network dynamical systems (i.e., the node set $\mathcal{V}$ and edge set $\mathcal{E}$ are constant over time). Although the network structure is thus not dynamic itself, its connectivity constrains how the node states $x_i(t)$ can influence each other. A crucial question in the study of network dynamical systems is therefore to characterise this influence of the network structure on the overall dynamics of the system. Vice versa, based on an observed network dynamics, we may also infer certain properties of the network.

A broad setup that is often considered in this context is the following set of autonomous, coupled ordinary differential equations,

$$\dot{x}_i = f_1(x_i) + \sum_{j=1,\ j\neq i}^{n} A_{ij} f_2(x_i, x_j), \quad \text{for all} \quad i \in 1, \dots, n, \tag{2.5}$$

in conjunction with an initial condition $\boldsymbol{x}(0) = \boldsymbol{x}_0$. Here the function f_1 describes the intrinsic dynamics of the node (a form of self-coupling), and the function f_2 describes how states of two nodes interact with each other (e.g., nodes are assumed to have pairwise interactions, in agreement with a network representation). Note how the adjacency matrix in Eq. (2.5) ensures that nodes that are not connected do not influence each other directly. These types of dynamical models may appear in a variety of contexts such as synchronisation, decentralised consensus, and social dynamics (Bullo, 2019).

A simple, but important case is linear dynamics on a network, in which case the above dynamics can be written in the form

$$\dot{\boldsymbol{x}} = \boldsymbol{F}\boldsymbol{x} \quad \text{with} \quad \boldsymbol{x}(0) = \boldsymbol{x}_0 \tag{2.6}$$

for some matrix $\boldsymbol{F}$, which we call the system matrix. We emphasise that given an undirected network with adjacency matrix $\boldsymbol{A}$, only certain linear dynamics will be compatible with graph structures encoded by $\boldsymbol{A}$ (see Section 2.3.2). Specifically, we will require that we can write $\boldsymbol{F} = \boldsymbol{D}\boldsymbol{A}_{\mathcal{G}}\boldsymbol{D}^{-1}$ for some invertible diagonal matrix $\boldsymbol{D}$ and an appropriately defined non-negative, symmetric matrix $\boldsymbol{A}_{\mathcal{G}} = \boldsymbol{A}_{\mathcal{G}}^\top$ that has the same sparsity pattern as the adjacency matrix for all off-diagonal entries. In particular, for $i \neq j$ we require that $[A_{\mathcal{G}}]_{ij} = F_{ij} = 0$ if $A_{ij} = 0$, and only nodes connected in the network interact directly.

Given such a system, linear systems theory tells us that the solution to the system just described is given by

$$\boldsymbol{x}(t) = \exp(t\boldsymbol{F})\boldsymbol{x}_0 = \left(\sum_{i=0}^{\infty} \frac{t^i \boldsymbol{F}^i}{i!}\right) \boldsymbol{x}_0, \tag{2.7}$$

where $\exp(\cdot)$ denotes the matrix exponential. In a discrete-time setting, the corresponding system of equations takes the form

$$\boldsymbol{x}(t+1) = \boldsymbol{G}\boldsymbol{x}(t) \quad \text{with} \quad \boldsymbol{x}(0) = \boldsymbol{x}_0, \tag{2.8}$$

with solution

$$\boldsymbol{x}(t) = \boldsymbol{G}^t \boldsymbol{x}_0, \tag{2.9}$$

where $\boldsymbol{G}^t$ denotes the tth power of the system matrix. We put the same restrictions on the system matrix $\boldsymbol{G}$ to be compatible with a network with adjacency matrix $\boldsymbol{A}$ as we did in the continuous time case (see Section 2.3.2).

The formal solutions (2.7) and (2.9) clearly show the importance of walks and indirect connectivity for linear processes on networks. As powers $\boldsymbol{A}^l$ of the binary adjacency matrix provide the number of walks of length l between any pair of nodes, powers of the system matrix encode weighted versions of these walks. More precisely, for a weighted network the entry $[\boldsymbol{A}^l]_{ij}$ will correspond to the weighted sum over all walks from i to j of length l, where the weight of a walk is given by the product of the weights of the edges it traverses. Walks provide the way for a node to spread its influence beyond its direct neighbours. Specifically, a node will be able to (indirectly) influence all the nodes within the same connected component, as there will exist a walk connecting any pair of nodes within each connected component. Connected components thus impose limitations on any dynamical process taking place on the network. Intuitively, a connected component is an island, and two nodes in different connected components cannot influence each other, even indirectly. In epidemic processes, for example, the existence of distinct components implies that certain regions of the network cannot be infected independently of the model of epidemic dynamics and its parameters, unless there was an initially infected node within the component. The notion of connectedness becomes more complicated in the case of directed networks, as there may be a path from node i to j but not vice versa.

2.3.2 Linear Network Dynamical Systems versus General Linear Systems

Why don't we allow for a generic linear system in the form of (2.6) or (2.8) when considering a linear dynamics on an undirected graph? For simplicity, let us explain the underlying issue here for the discrete time setup, as the continuous time case is analogous. Clearly, any matrix $\boldsymbol{G} \in \mathbb{R}_+^{n \times n}$ induces a weighted (directed) graph if we consider the node set $\mathcal{V}$ to correspond to the integers $1, \ldots, n$, and the edge set $\mathcal{E}$ to correspond to the nonzero entries in $\boldsymbol{G}$. More precisely, we can define $\mathcal{E} = \{(i, j) | G_{ij} \neq 0\}$, with an accordingly defined weight $w_{ij} = G_{ij}$ for each edge. We could thus argue that since the underlying graph was supposed to be undirected, the requirement that the system matrix $\boldsymbol{G}$ needs to fulfill is that it should be symmetric.

Associating *any* matrix with a graph, however, is not fruitful in a general dynamical context. The reason is somewhat subtle and intrinsically linked to the interpretation of the state variables x_i rather than the matrix $\boldsymbol{G}$. Consider a generic linear system where the states x_i have no specific meaning assigned to them but are simply arbitrary coordinates in which we measure the state of our system. For simplicity, let us assume that these coordinates evolve according to the discrete time system $\boldsymbol{x}(t+1) = \boldsymbol{A}\boldsymbol{x}(t)$, where $\boldsymbol{A}$ can be interpreted as the adjacency matrix of some graph $\mathcal{G}$. As the coordinates we chose are arbitrary, we may as well use a different set of coordinates to measure our system state. For instance, we could choose the vector $\boldsymbol{y} = \boldsymbol{Z}\boldsymbol{x}$ as our new state vector, for some invertible matrix $\boldsymbol{Z}$. While the dynamics of the system has not changed – we are simply measuring it in a different reference system – the system matrix will transform, and we will have a discrete time system of the form $\boldsymbol{y}(t+1) = \boldsymbol{Z}\boldsymbol{A}\boldsymbol{Z}^{-1}\boldsymbol{y}(t)$. But we can, of course, interpret $\boldsymbol{Z}\boldsymbol{A}\boldsymbol{Z}^{-1} = \boldsymbol{A}'$ as the adjacency matrix of another graph $\mathcal{G}'$. In fact, as we assumed a symmetric $\boldsymbol{A}$, we could choose a matrix $\boldsymbol{Z}$ that diagonalizes the adjacency matrix $\boldsymbol{A}$. With this choice we would arrive at a new matrix $\boldsymbol{A}'$ with a diagonal coupling, with eigenvalues of the original matrix on the diagonal. However, the matrix $\boldsymbol{A}'$ would correspond to a graph with no edges between different nodes at all.

As this example illustrates, associating a graph to any matrix can be problematic when interpreted in the context of a dynamical system. Recall that when modelling a system as a network, we associate a dynamical state (or a state vector) to each node, and we assume that direct interactions between node states happen only over the edges in the graph. A key role is thus played by the state variables x_i. The example shows that we cannot change the state variable in an arbitrary way if we want our state variables to remain interpretable as some local information on a specific node. For instance, consider a social network, where we associate a scalar opinion variable to each node and consider a linear opinion formation process. We could potentially diagonalise the system to bring it into a simpler algebraic form, and we may actually do so to gain insight into the long-time behaviour of the dynamics. However, the new state variables of the diagonalised system then correspond to weighted sums over all previous state variables in general (i.e., to a summation over all nodes). Hence, the new state variables can clearly not be associated with a single node in the original network, but incorporate information that is distributed across the whole original network. Accordingly, the diagonalised system described in the new state variables does not describe a distributed dynamical system. Thus a general system matrix cannot be interpreted as a network dynamical system, despite the fact that the interaction pattern encoded in the system matrix may be sparse.

The only types of state transformation that are guaranteed to leave the state variables in a network localised at their original nodes for Eq. (2.6) and (2.8) are those corresponding to a linear scaling (and a measurement offset), that is, $y_i = a_i x_i + b_i$ for scalar node states, which will be our primary focus in the following discussion. As a measurement offset will simply translate into an external input in the new state variables, we will consider only the case $b_i = 0$. This scaling of the node variables then corresponds precisely to the requirement that $\boldsymbol{G} = \boldsymbol{D}\boldsymbol{A}_{\mathcal{G}}\boldsymbol{D}^{-1}$ (and similarly $\boldsymbol{F}$) should be related by a diagonal similarity transformation to a matrix whose (off-diagonal) sparsity pattern is commensurate with the graph's adjacency matrix.

2.3.3 Spectral Decomposition

As our discussion has emphasised so far, we should exercise care when performing coordinate transformations of the state vectors. However, the solutions (2.7) and (2.9) take a particularly simple form when changing our state coordinates to the eigenmodes of the system, as we will illustrate for continuous time dynamics below (the discrete case is analogous). For simplicity, we will continue to assume that the dynamics takes place on an undirected graph.

In the continuous-time setting, assuming $\boldsymbol{F} = \boldsymbol{D}\boldsymbol{A}_{\mathcal{G}}\boldsymbol{D}^{-1}$, then the system matrix has real eigenvalues λ_i such that (i) $\boldsymbol{F}\boldsymbol{v}_i = \lambda_i \boldsymbol{v}_i$ for some right eigenvector $\boldsymbol{v}_i$, (ii) $\boldsymbol{u}_i^\top \boldsymbol{F} = \lambda_i \boldsymbol{u}_i^\top$ for some left eigenvector $\boldsymbol{u}_i$, (iii) $\boldsymbol{u}_i^\top \boldsymbol{v}_i = \delta_{ij}$, where δ_{ij} is the Kronecker delta, and (iv) $\boldsymbol{v}_i = \boldsymbol{D}^2 \boldsymbol{u}_i$. Assuming further that $\boldsymbol{F}$ is a (marginally) stable system, we can order these eigenvalues such that $0 \geq \lambda_1 \geq \lambda_2 \geq \cdots \geq \lambda_n$. We now expand $\boldsymbol{F}$ using this bi-orthogonal spectral expansion:

$$\boldsymbol{F} = \sum_{i=1}^{n} \lambda_i \boldsymbol{v}_i \boldsymbol{u}_i^\top. \tag{2.10}$$

Accordingly, the solution to our linear dynamics (Eq. (2.6)) on the network can be written as

$$\boldsymbol{x}(t) = \sum_{i=1}^{n} e^{\lambda_i t}\, \boldsymbol{v}_i \boldsymbol{u}_i^\top \boldsymbol{x}_0 = \sum_{i=1}^{n} \left(\boldsymbol{u}_i^\top \boldsymbol{x}_0\right) e^{\lambda_i t}\, \boldsymbol{v}_i. \tag{2.11}$$

In other words, each of the eigenvectors of the system matrix defines a mode, and the exponential decay of each mode is determined from its corresponding eigenvalue. The dynamics is thus entirely determined by the eigenvectors and eigenvalues of that matrix. This is, of course, a well-known result for linear systems. From a network dynamics perspective, the following question is nonetheless of interest: how do the spectral properties of the system matrix, and hence the linear dynamics, depend on the structural properties of a graph? This question is at the core of spectral graph theory, which considers precisely this issue for matrices such as the adjacency matrix and the graph Laplacian.

THE CASE OF DIRECTED NETWORKS

For the sake of simplicity, most of the technical derivations in this Element assume that the network under consideration is undirected and, accordingly, its adjacency matrix will be symmetric, so that its eigenvectors form an orthonormal basis. In situations when the matrix is asymmetric and non-normal, the transformation to eigenvector coordinates may involve a strong distortion of the vector space and induce non-intuitive spectral properties (Trefethen & Embree, 2005). By non-normal, we mean a matrix $\boldsymbol{A}$ such that $\boldsymbol{A}\boldsymbol{A}^T \neq \boldsymbol{A}^T\boldsymbol{A}$, which can occur only for asymmetric matrices. Within the language of networks, this condition means that at least two nodes have different numbers of out-neighbours and of in-neighbours. Non-normality can lead to unexpected dynamical patterns for a linear system. For instance, the system can undergo a transient growth before asymptotically converging to zero, as measured by the norm of the state vector $\mathbf{x}$, even if the real part of its eigenvalues are all negative. This transient behaviour is not explained by the eigenvalues of the matrix $\boldsymbol{A}$ and can have important consequences, especially when the linear dynamics is an approximation of a non-linear dynamical system. The non-normality of a matrix can be quantified by the so-called Henrici's departure from normality. Non-normal adjacency matrices have been observed empirically in a wide range of directed networks (Asllani, Lambiotte, & Carletti, 2018).

2.4 Laplacians, Diffusion and Consensus

Linear dynamics on networks like Eq. (2.6) and (2.8) appear in many contexts. For instance, we may arrive at such systems by linearising a nonlinear dynamical system on a network around a fixed point. In this case the system matrix will be the Jacobian matrix.[4] Another important class of linear models is diffusion (linear Markov chains) and consensus dynamics. In this section, we will

[4] Note that the Jacobian is a matrix depending not only on the underlying network structure but also on the details of the non-linear dynamical model as well as of the fixed point around which the linearisation is performed. Importantly, there is no guarantee that the Jacobian is a non-negative matrix, and thus an interpretation in terms of a network dynamical system may or may not be fruitful (see Section 2.3.2).

give a short overview of these models and introduce the associated Laplacian matrices that will appear pervasively throughout this Element.

Diffusion is a central concept in almost any field of science. In its simplest setting, where an entity diffuses in a continuous environment, its dynamics is determined by the diffusion equation, also known as the heat equation, a partial differential equation first studied by Joseph Fourier to model how heat diffuses in a physical medium:

$$\partial_t \rho(\mathbf{x}, t) = \mathbf{\Delta} \rho(\mathbf{x}, t). \tag{2.12}$$

Here $\mathbf{\Delta}$ is the Laplacian (e.g., $\mathbf{\Delta} = \frac{\partial^2}{\partial x^2} + \frac{\partial^2}{\partial y^2}$ for Cartesian coordinates in two dimensions, and $\rho(\mathbf{x}, t)$ represents the density of particles or energy at a certain position $\mathbf{x}$ at time t). The diffusion equation is without hesitation one of the most widely studied equations in applied mathematics, and a lot is known about its solutions with different boundary conditions, as well as its many applications, from biology to finance. The diffusion equation and the Laplacian naturally emerge in a variety of stochastic models (sometimes as a particular limit). When defining an analogue of the diffusion equation for networks, there are numerous possible choices to make, leading to somewhat different types of Laplacian matrices that can be defined for the same graph.

2.4.1 The Combinatorial Graph Laplacian

The combinatorial Laplacian matrix $\boldsymbol{L}$ of a graph is the $n \times n$ matrix with entries

$$L_{ij} = \begin{cases} k_i = \sum_j A_{ij}, & \text{if } i = j, \\ -A_{ij}, & \text{if } \{i, j\} \in E, \\ 0, & \text{otherwise.} \end{cases} \tag{2.13}$$

The combinatorial Laplacian naturally appears in diffusion and consensus problems, where the state of each node evolves towards the states of its neighbours according to

$$\frac{d}{dt} x_i = \sum_j A_{ij}(x_j - x_i), \tag{2.14}$$

which can be rewritten in matrix form as

$$\frac{d}{dt} \boldsymbol{x} = -\boldsymbol{L}\, \boldsymbol{x}, \tag{2.15}$$

where $\boldsymbol{L} = \boldsymbol{K} - \boldsymbol{A}$, and $\boldsymbol{K}(\boldsymbol{A}) = \text{diag}(\boldsymbol{A}\mathbf{1})$ is the diagonal matrix of node degrees.

Equation (2.15) can be seen as a discrete analogue of the diffusion equation (Eq. (2.12)).[5] The name consensus dynamics derives from the fact that, as $t \to \infty$, the dynamics will converge to the average of the initial node states (i.e., $\lim_{t\to\infty} \boldsymbol{x}(t) = \mathbf{1}(\mathbf{1}^\top \boldsymbol{x}_0)/n$), if the graph is connected, which we will be considering from now on unless stated otherwise. Thus, for large times $t \to \infty$, we will see a homogenisation of the states of the nodes, analogously to how we will see an equilibration of temperature in the heat equation. In analogy with how Fourier modes are the eigenfunctions of the continuous Laplace operator, using eigenvectors of the combinatorial Laplacian as a unitary transformation is sometimes called a graph Fourier transform (Shuman et al., 2013); and, as we shall see later, akin to the Fourier transform we can similarly ascribe interpretations of 'slow' and 'fast' eigenmodes to the eigenvectors, depending on the associated eigenvalue.

If the underlying graph is undirected, it is easy to show that the Laplacian is symmetric and positive semidefinite. This can be proven by noting that the Laplacian naturally defines a quadratic form,

$$\boldsymbol{x}^\top \boldsymbol{L} \boldsymbol{x} = \frac{1}{2}\sum_{i,j}^{n} A_{ij}(x_i - x_j)^2, \tag{2.16}$$

from which we can further deduce that the Laplacian has one eigenvector given by $\mathbf{1} = (1, \ldots, 1)^\top$ associated to eigenvalue 0. In general, there will exist one eigenvector with eigenvalue 0 for each connected component of the graph (i.e., the multiplicity of the zero eigenvalue is equal to the number of connected components of the graph). Hence, the combinatorial Laplacian has real, non-negative eigenvalues, usually ordered as $\lambda_1 = 0 \leq \lambda_2 \leq \cdots \leq \lambda_n$. Finally, we remark that the combinatorial Laplacian is not affected by the addition of self-loops to nodes.

2.4.2 The Normalised and the Random-Walk Graph Laplacian

The normalised graph Laplacian is the $n \times n$ matrix defined via

$$\mathcal{L}_{ij} = \begin{cases} 1, & \text{if } i = j, \\ -A_{ij}/\sqrt{k_i k_j}, & \text{if } \{i, j\} \in E, \\ 0, & \text{otherwise.} \end{cases} \tag{2.17}$$

In matrix form we have $\boldsymbol{\mathcal{L}} = \boldsymbol{I} - \boldsymbol{K}^{-1/2}\boldsymbol{A}\boldsymbol{K}^{-1/2}$. The normalised Laplacian exhibits properties similar to those of the combinatorial Laplacian. The multiplicity of the eigenvalue 0 indicates the number of connected components, and

[5] Note that due to convention, there is a difference in the sign of those two operators.

the matrix is positive semi-definite, with eigenvalues $\mu_1 = 0 \leq \mu_2 \leq ... \leq \mu_n \leq 2$. Note that for regular graphs, the normalised and combinatorial Laplacians are equivalent up to a multiplicative factor (i.e., for graphs in which all nodes have the same degree). Furthermore, we will have $\mu_n = 2$ only if the graph is bipartite.

While the normalised Laplacian is commonly analysed in spectral graph theory, in the context of network dynamics the closely related random-walk Laplacian $\boldsymbol{L}_{\text{rw}}(\boldsymbol{A}) = \boldsymbol{I} - \boldsymbol{K}^{-1}\boldsymbol{A}$ is often the more relevant matrix, as it appears naturally in the context of continuous time diffusion processes on graphs. In terms of its entries, the random-walk Laplacian is written as

$$(\boldsymbol{L}_{\text{rw}})_{ij} = \begin{cases} 1, & \text{if } i = j, \\ -A_{ij}/k_i, & \text{if } \{i, j\} \in E, \\ 0, & \text{otherwise.} \end{cases} \tag{2.18}$$

While the random-walk Laplacian is asymmetric, its spectrum can be directly obtained from the normalised Laplacian, as both matrices are related by the similarity transformation $\boldsymbol{L}_{\text{rw}} = \boldsymbol{K}^{1/2}\boldsymbol{\mathcal{L}}\boldsymbol{K}^{-1/2}$. It thus follows that the random-walk Laplacian $\boldsymbol{L}_{\text{rw}}$ and normalised Laplacian $\boldsymbol{\mathcal{L}}$ have exactly the same real spectrum. Note that this similarity transformation is precisely of the type discussed in Section 2.3.2. Further, the left and right eigenvectors $\boldsymbol{u}$ and $\boldsymbol{v}$ of $\boldsymbol{L}_{\text{rw}}$ are related to the eigenvectors $\boldsymbol{w}$ of $\boldsymbol{\mathcal{L}}$ via $\boldsymbol{v}_i = \boldsymbol{K}^{1/2}\boldsymbol{w}_i$ and $\boldsymbol{u}_i = \boldsymbol{K}^{-1/2}\boldsymbol{w}_i$ (cf. Section 2.3.2).

Similar to the combinatorial Laplacian, the random-walk Laplacian can also be associated to a (weighted) consensus dynamics,

$$\frac{d}{dt}x_i = \frac{1}{k_i}\sum_j A_{ij}(x_j - x_i), \tag{2.19}$$

or, in matrix form,

$$\frac{d}{dt}\boldsymbol{x} = -\boldsymbol{L}_{\text{rw}}\,\boldsymbol{x}. \tag{2.20}$$

In contrast to the 'standard' consensus dynamics (Eq. (2.15)), the rate at which the nodes adapt their states is now modulated by their degrees. This leads ultimately to a weighted consensus value $\lim_{t\to\infty}\boldsymbol{x}(t) = \boldsymbol{1}(\boldsymbol{k}^\top\boldsymbol{x}_0)/(\boldsymbol{k}^\top\boldsymbol{1})$, where $\boldsymbol{k}$ is the vector of node degrees.

As the name suggests, however, the random-walk Laplacian is typically considered in the context of a diffusion, modelled via a random-walk process,

$$\frac{d}{dt}\boldsymbol{p}^\top = -\boldsymbol{p}^\top\boldsymbol{L}_{\text{rw}}, \tag{2.21}$$

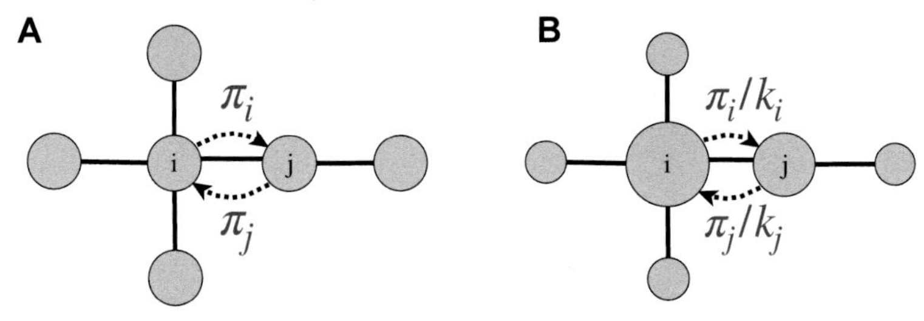

Figure 2 Dominant eigenvectors of Laplacian matrices. The stationary states of Laplacian processes on undirected networks exhibit detailed balance: the flow along any edge from i to j is equal to the flow from j to i. For the combinatorial Laplacian, the stationary state is uniform on the nodes. For the random-walk Laplacian, in contrast, it is proportional to the degree of the nodes.

where $\boldsymbol{p}^\top$ is a row vector encoding the probability that a random walker is located at a particular node at time t. This process can be seen as a continuous-time version of a random walk defined on the nodes of the network where, at each time step, a walker located on a node follows an outgoing edge at random to another node with a probability proportional to the outgoing edge weight. The transition probability between nodes i and j can thus be written as

$$T_{ij} = \frac{A_{ij}}{k_i}, \tag{2.22}$$

and the associated discrete-time random walk process can be compactly described via the linear difference equation:

$$\boldsymbol{p}^\top(t+1) = \boldsymbol{p}^\top(t)\boldsymbol{T}. \tag{2.23}$$

Note that Eq. (2.23) can be obtained by approximating $d\boldsymbol{p}^\top/dt$ in Eq. (2.21) by a discrete difference. From the spectral properties of the normalised Laplacian, it is clear that the process (Eq. (2.21)) is ergodic when the underlying network is undirected and connected. In that case, irrespective of the initial condition, the density of walkers converges to a unique stationary state $\lim_{t\to\infty} \boldsymbol{p}^\top(t) = \boldsymbol{\pi}^\top$. The probability of finding a walker on a particular node i at this stationary state is proportional to the node degree (i.e., $\pi_i = k_i/2m$, where $2m = \sum_i k_i$ is twice the number of edges (see Figure 2)). Note that the stationarity state of random-walk processes is also closely related to the PageRank of the nodes (Gleich, 2015; Langville & Meyer, 2011), a well-known centrality measure calculated recursively, based on the idea that an important node receives connections from many other important nodes.

Before concluding this subsection, let us note that the combinatorial Laplacian and the random-walk Laplacian are two of the most popular choices of Laplacian, even if other options exist. Both matrices have the same dominant right eigenvector, **1**, associated to the conservation of the number of walkers in the process but have, as we have seen, different dominant left eigenvectors. For instance, only the combinatorial Laplacian satisfies Fick's condition (Putra, Thompson, & Goriely, 2021), which states there is no net flux in the absence of 'concentration' gradient, so that the uniform state $\mathbf{1}^\top$ is the stationary solution for the diffusive process.[6] Different Laplacians may lead to different qualitative and quantitative dynamical patterns on the same underlying graph, and a specific choice must thus be made carefully to capture the important properties of the process that one wants to model. The combinatorial Laplacian is often preferred for physical and electrical networks, while the random-walk Laplacian is preferred instead for the diffusion of information in socio-economic systems.

2.5 Distances, Similarities, and Kernels

Several problems in machine learning, data mining, and network science require the calculation of a similarity or dissimilarity (distance) matrix between objects. Here, a dissimilarity between two objects i and j is a positive function $d(i,j) \geq 0$ that is symmetric, $d(i,j) = d(j,i)$, and such that $d(i,j) = 0$ if and only if $i = j$. Such a dissimilarity, which is more formally known as a semi-metric, is intuitively defined such that higher values indicate a higher difference between the objects. If, in addition, a semi-metric satisfies the so-called triangle inequality $d(i,k) \leq d(i,j) + d(j,k)$ for all i, j, k, then the function d is called a metric or distance function. If the function d satisfies the triangle inequality but can also be zero for different nodes, that is, $d(i,j) = 0$ for some $i \neq j$ (in addition to $d(i,i) = 0$), then d is called a pseudometric. In the context of network analysis, a dissimilarity between two nodes captures a certain notion of difference between them. Any dissimilarity measure can be transformed into a similarity measure by basic algebraic operations such that the resulting similarity is high when the two objects are close to each other.

[6] To be more precise, we define the concentration of a diffusion process on a node as the stationary probability of the random walk divided by the volume of the node. In many practical applications, all the nodes are assumed to have the same volume, and Fick's condition holds for the combinatorial Laplacian. However, other notions of volume have been proposed in graph theory; see our discussion on the Cheeger inequality in Section 4.4.1 for an example. If the volume is taken to be proportional to the node degree, then Fick's condition holds for the random-walk Laplacian instead. In essence, each Laplacian defines a different notion of volume for the nodes, which has an impact on community detection for instance, as we will discuss in Section 6.4.

2.5.1 Distance Measures on Graphs

A popular choice of dissimilarity between nodes i and j is the length of the shortest path between them, which can be shown to be a proper distance measure. Shortest paths are of particular relevance if we want to navigate between two nodes in a graph (e.g., in routing problems). However, the shortest path distance is very sensitive to inaccuracies in the network, as the addition or removal of edges may radically alter the shortest paths between nodes. Furthermore, the shortest path distance does not account for the possible presence of multiple, complementary paths connecting pairs of nodes.

An alternative distance measure between nodes is the average commute time of an unbiased random walk on the network. This quantity is also referred to as commute time distance and can also be shown to be equivalent to the so-called effective resistance between two nodes up to a scaling factor (Chandra et al., 1996). Here the *effective resistance* ω_{ij} is defined via the quadratic form

$$\omega_{ij} \triangleq (\boldsymbol{e}_i - \boldsymbol{e}_j)^T \boldsymbol{L}^\dagger (\boldsymbol{e}_i - \boldsymbol{e}_j), \tag{2.24}$$

where $\boldsymbol{L}^\dagger$ is the pseudoinverse of the (combinatorial) graph Laplacian, defined as the inverse of $\boldsymbol{L}$ in the space orthogonal to its kernel; that is,

$$\boldsymbol{L}^\dagger = \sum_{i=2}^{n} \frac{1}{\lambda_i} \boldsymbol{u}_i \boldsymbol{u}_i^\top. \tag{2.25}$$

$\boldsymbol{e}_i$ is the indicator vector of node i, which has a value of 1 at position i and is zero otherwise. Originally defined in the context of electrical circuit theory, the effective resistance has found its way into graph theory-related areas through various applications, including graph embeddings (Fiedler, 2011) and, more recently, graph sparsification (Spielman & Teng, 2011). Importantly, the effective resistance is also a distance metric (Gvishiani & Gurvich, 1987; Klein & Randić, 1993), and a small effective resistance between two nodes indicates their proximity in the graph. In particular, the effective resistance distance decreases when the number of paths connecting two nodes increases and when the lengths of these paths decreases (Devriendt, 2020). Note that the effective resistance between any two nodes is equal to their shortest-path distance for a tree graph. Due to its connection to the commute time, the resistance distance is a natural choice of dissimilarity measure for networks in which follow the do not necessarily follow shortest paths (e.g., in the context of information diffusion or virus spreading).

2.5.2 Similarities and Kernels

Within the family of similarity measures, kernels, also called kernel functions, have the additional property that they are obtained from the inner product between representations of the nodes in a vector space; that is,

$$\kappa(i,j) = \boldsymbol{v}^{\top}(i)\boldsymbol{v}(j), \tag{2.26}$$

where $\boldsymbol{v}(i)$ is the vector representation of node i (see Section 7.1 for a longer discussion on network embeddings). Importantly, the embedding of the nodes does not need to be computed explicitly, as the kernel is the result of the inner product between the vectors, a property usually called *the kernel trick* (Fouss, Saerens, & Shimbo, 2016). From its definition in terms of inner products, a kernel function is symmetric and positive semi-definite. Besides these mathematical properties, the formulation in terms of an inner product gives an intuitive interpretation. Important examples for graph kernels include the different forms of exponential kernels (Kondor & Lafferty, 2002), such as the diffusion, or heat, kernel (cf. Eq. (2.15)):

$$\kappa(i,j) = \left(e^{-t\boldsymbol{L}}\right)_{ij}. \tag{2.27}$$

From the spectral decomposition of the Laplacian, this kernel matrix can be equivalently constructed from the inner products of the following vectors associated to each node i

$$\boldsymbol{v}(i) = \left([\boldsymbol{u}_1]_i,\ e^{-t\lambda_2/2}[\boldsymbol{u}_2]_i,\ \ldots,\ e^{-t\lambda_n/2}[\boldsymbol{u}_n]_i\right), \tag{2.28}$$

where $[\boldsymbol{u}_l]_i$ is the ith entry of the lth eigenvector of the Laplacian $\boldsymbol{L}$, λ_l is its corresponding eigenvalue of $\boldsymbol{L}$, and we have used the fact that $\lambda_1 = 0$. Note that other system matrices can be chosen instead of the Laplacian matrix. If the adjacency matrix is chosen, the kernel is equivalent to the so-called communicability matrix $e^{\boldsymbol{A}}$ (Estrada & Hatano, 2008) when $t = 1$. In general, for any symmetric matrix $\boldsymbol{S}$ the exponential of that matrix is a positive semi-definite kernel, as the eigenvalues of the matrix exponential will correspond to the exponential function applied to the original eigenvalues (Higham, 2008).

2.6 Further Discussion and References

Network science is a rich field of research, and we have only presented a limited selection of results and concepts in this section, whose understanding is critical for the rest of this Element. For readers who search more thorough presentations of the field, we refer to introductory books by Newman (2018a), Barabási et al. (2016), or Menczer, Fortunato, and Davis (2020). Monographs or review articles on more specialised and advanced aspects of Network Science include F. R. Chung (1997) for spectral graph theory, Strogatz (2004) and Arenas et al. (2008) for the non-linear dynamics of synchronisation on networks, Fouss et al. (2016) for similarity measures and kernels on networks, and Masuda et al. (2017) for random walks on networks.

3 Modularity, Community Detection, and Clustering in Networks

What does it mean for a network to have a modular structure or consist of several communities? In this section we discuss three common ways to conceptualise modular network structure and detect such structure in networks.

3.1 Communities as Clusters: Modularity and Assortative Communities

An essential problem in Data Science is clustering (Xu & Wunsch, 2008): the unsupervised partitioning of objects into groups, such that objects within the same cluster are more similar to each other (in a certain sense) than to objects in another cluster. Though clustering is typically considered for points in a metric space, the idea of clustering can be generalised to networks as well. Analogously to the clustering of data points, we may informally describe the problem as the partitioning of nodes into groups, such that nodes within each group are more similar to each other than to nodes outside their group.[7] In the context of networks, this problem is typically called community detection (Fortunato, 2010; Fortunato & Hric, 2016; Schaeffer, 2007; Schaub et al., 2017). As we will discuss extensively, community detection finds important implications to understand dynamics, but other critical applications include network visualisation (Bohlin et al., 2014; Komarek, Pavlik, & Sobeslav, 2015), task optimisation (Bui-Xuan & Jones, 2014), and node classification (Cavallari et al., 2017).

Community detection is often formalised as follows: first, a quality function is defined that assigns a score for each possible network partition. Second, an optimisation procedure is employed in order to find the partition optimising the quality function. As most problem formulations of community detection involve difficult combinatorial problems, heuristic optimisation procedures are often employed in practice. Importantly, the number of groups in the network is typically assumed to be unknown and has to be inferred from the network data as well (Fortunato, 2010).

One of the most popular quality functions for community detection is the so-called Newman–Girvan modularity (Newman & Girvan, 2004), denoted by Q. Let us consider a group of nodes defined by a set $\mathcal{A}$. The underlying idea of modularity is to compare the number of links connecting nodes inside $\mathcal{A}$ with an expectation of this number under a random null model. The choice

[7] In this Element, we will exclusively focus on community structure made of non-overlapping groups. Note that methods have been specifically designed for the detection and characterisation of overlapping communities, as in Ahn, Bagrow, and Lehmann (2010).

of null model is, in principle, left to the user. The null model should ideally be tailored to the type of network under scrutiny and the existence of forces that may constrain the formation of edges (Expert et al., 2011). However, the default choice is often the soft configuration or Chung–Lu model (F.Chung & Lu, 2002), for which the expected weight of an edge between two nodes i and j is given by Eq. (2.3).

Under the soft-configuration model, the difference between the number of links in community $\mathcal{A}$ and the expected value of such links is

$$\sum_{i,j\in\mathcal{A}} \left(A_{ij} - \frac{k_i k_j}{2m}\right), \tag{3.1}$$

where each term may be either positive or negative depending on the presence or absence of a link between two nodes. Note that the null model penalises missing links between pairs of higher-degree nodes more than missing links between nodes with small degree. The Newman–Girvan modularity now equates the quality of a partition $\mathcal{P} = \{\mathcal{A}_1, \mathcal{A}_2, \ldots, \mathcal{A}_C\}$ to the sum over the contributions of type (3.1) of every community:

$$Q = \frac{1}{2m} \sum_{\mathcal{A}_\alpha \in \mathcal{P}} \sum_{i,j\in\mathcal{A}_\alpha} \left(A_{ij} - \frac{k_i k_j}{2m}\right), \tag{3.2}$$

where the prefactor $1/(2m)$ ensures that modularity is smaller than 1 in absolute value. Intuitively, the modularity measure thus assigns high scores to communities if they are densely connected internally, but only weakly connected to other communities. This form of community structure is often called *assortative* community structure, in contrast to the *disassortative* and more general *block* structures presented in Section 3.3.

Let us encode the partition of the graph into C communities with the $N \times C$ indicator matrix $\boldsymbol{H}$, where $H_{i\alpha}$ is equal to 1 if node i belongs to community $\mathcal{A}_\alpha$, and zero otherwise. The modularity of a partition associated to $\boldsymbol{H}$ can then be written in matrix notations as follows:

$$Q = \frac{1}{2m} \mathrm{Tr}\left[\boldsymbol{H}^T \left[\boldsymbol{A} - \frac{\boldsymbol{k}\boldsymbol{k}^\top}{2m}\right] \boldsymbol{H}\right], \tag{3.3}$$

where $\boldsymbol{k}$ is the vector of node degrees and Tr denotes the trace of a matrix.

Inherent to the construction of modularity is the assumption that a network partition with a strong assortative community structure will lead to high values of modularity, in the sense that an unexpectedly large number of edges will be concentrated inside its communities. Modularity optimisation (i.e., finding the partition of a network having the highest value of modularity) has thus been proposed as one way to solve the community detection problem. As modularity optimisation is NP-hard (Brandes et al., 2007), several heuristics have been

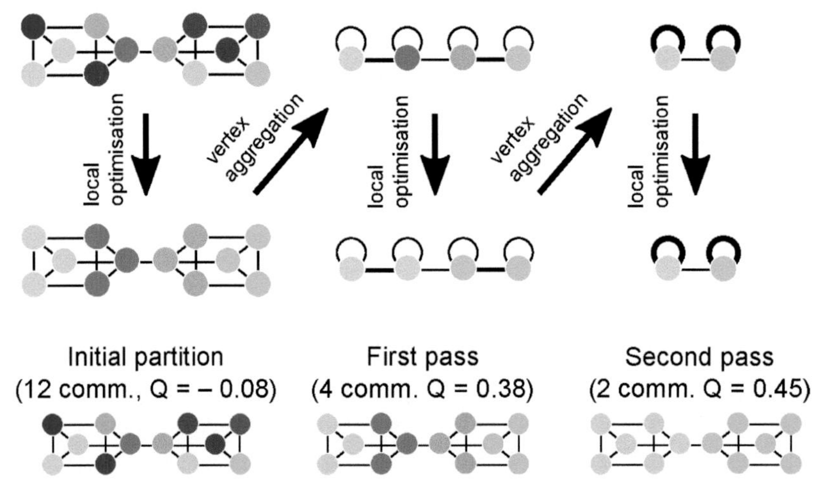

Figure 3 Louvain method for modularity optimisation. The Louvain method consists of repeatedly applied passes over the network until no further increase in modularity is observed. Each pass is split into two phases. The first phase consists in a local optimisation, where each vertex can be moved to the community of its direct neighbours. The second phase aggregates vertices and constructs a meta-graph whose vertices are the communities found after the first phase. Figure adapted from Aynaud et al., (2013).

proposed for modularity optimisation. Some of the most popular techniques include (i) spectral methods, where the communities are uncovered from the eigenvectors of a matrix describing the graph (Newman, 2013; Von Luxburg, 2007), and (ii) greedy methods, proceeding by agglomerating groups of nodes into larger ones to improve modularity (Fortunato, 2010).

In the following, we discuss the Louvain method (Blondel et al., 2008), a greedy method that has become a standard choice for modularity optimsation due to its good balance between simplicity, accuracy, and speed. The algorithm starts with a weighted graph where the n vertices are randomly assigned an index between 0 and $n-1$. It consists of two phases that are repeated iteratively until a local maximum of modularity is reached, as illustrated in Figure 3.

1. The first phase works as follows. An initial partition is constructed by placing each vertex into a separate community (e.g., each community is composed of one single node). We then consider the first vertex, with index 0, and calculate the change of modularity by removing it from its community and placing it into the community of one of its neighbours. This potential change is calculated for each of the neighbours of 0. The vertex 0 is then moved to the community with maximal positive increase. If

there is no community with a positive increase, the node is placed back into its original community. This process is applied sequentially to all the vertices. After applying it to node $n - 1$, one returns to node 0 and iterates until no vertex is moved during a complete iteration of n steps. The first phase is then finished, and its outcome is a partition of the network into C communities.

2. The second phase builds an aggregated weighted graph. The nodes in this new aggregated graph correspond to the C communities discovered during the first phase, and the weight of the links between two such nodes is given by the sum of the link-weights between the nodes in the original network in these two communities. The links that existed between the vertices of the same community create loops in the new aggregated graph.

Note how the first phase finds a local optimum of modularity, where the search is done locally by moving nodes to the communities of their direct neighbours. The second phase then changes the scale over which the optimisation is performed, by allowing movement of groups of nodes to improve modularity, instead of single nodes. This two-phase approach searches for structures in a multiscale way, reminiscent of the self-similarity often observed in complex systems (Serrano, Krioukov, & Boguná, 2008; Simon, 1962). Both phases are repeated until no further increase in modularity is found. The Louvain algorithm is especially efficient because the change in modularity when moving a node to a community of one of its neighbours can be computed fast, based only on local information around the node that is to be moved.

Despite their popularity, modularity maximisation methods are not exempt from limitations. Modularity suffers from a so-called resolution limit which makes it impossible to detect communities of nodes that are smaller than a certain scale (Fortunato & Barthelemy, 2007). In other words, even if this is not apparent from the definition of modularity at first sight, modularity tends to favour partitions where the communities have a characteristic size depending on the total size of the system. The resolution limit arises from the dependency of the null model $k_i k_j / 2m$ on the total weight $2m$ (number) of the edges. This dependency makes the negative term in modularity smaller when the total weight (number) of edges is larger, and thus favours communities made of more nodes if $2m$ increases. Another limitation of modularity is that its landscape over the space of partitions is usually extremely rugged (Good, De Montjoye, & Clauset, 2010), with multiple local maxima close to the global optimum, which may limit the interpretability of the approximate solutions found by modularity optimisation. Finally, the Louvain algorithm may in certain cases lead to intermediate disconnected communities, which cannot be optimal for modularity. This latter problem can, however, be remedied by adjusting the Louvain algorithm accordingly (Traag, Waltman, & Van Eck, 2019).

3.2 Communities Defined via Sparse Cuts: Graph Partitioning and Spectral Methods

Thinking about communities in terms of sets of densely connected nodes, as done when considering modularity, is one important perspective on community detection. However, many graph partitioning methods employ a different perspective (Schaub et al., 2017). Instead of searching for groups with a large number of edges inside, they aim instead to find a minimal set of cuts in the graph, such that the resulting node groups have a balanced size according to some criterion (Von Luxburg, 2007). This type of formulation was probably first considered in circuit layout, where one is confronted with a graph which describes the signal flows between different components of a circuit (Alpert & Kahng, 1995). However, since then, this type of graph partitioning has been employed in many other contexts as well (Von Luxburg, 2007).

Let us consider the simplest instance of the problem of finding the best bipartition of a network such that the number of edges between two groups $\mathcal{A}_\alpha$ and $\mathcal{A}_\beta$ is minimised, following Newman (2013). The cut size R of a partition counts the number of edges existing between the two groups of vertices and can be written as

$$R = \frac{1}{2} \sum_{\substack{i,j \text{ in} \\ \text{different} \\ \text{groups}}} A_{ij}, \tag{3.4}$$

where we divide by 2 as each edge is counted twice when summing over the node indices. This quantity can be rewritten more conveniently by defining the indices:

$$s_i = \begin{cases} 1 & \text{if node } i \text{ is in group } \mathcal{A}_\alpha, \\ -1 & \text{if node } i \text{ is in group } \mathcal{A}_\beta. \end{cases} \tag{3.5}$$

Using the fact that

$$\frac{1}{2}(1 - s_i s_j) = \begin{cases} 1 & \text{if nodes } i \text{ and } j \text{ are in different groups,} \\ 0 & \text{if nodes } i \text{ and } j \text{ are in the same group,} \end{cases} \tag{3.6}$$

and some algebra, we rewrite the cut size R in terms of the Laplacian matrix $\boldsymbol{L}$ as

$$R = \frac{1}{4} \sum_{ij} s_i L_{ij} s_j, \tag{3.7}$$

or equivalently in matrix form:

$$R = \frac{1}{4}\mathbf{s}^T L\mathbf{s}. \tag{3.8}$$

Hence, finding the minimal cut is equivalent to choosing the vector $\boldsymbol{s}$ that minimises Eq. (3.8). This expression already appeared, as Eq. (2.16), when we showed that the Laplacian is positive semi-definite and the vector **1** is an eigenvector of the Laplacian with eigenvalue 0. Therefore, from the properties of the Laplacian matrix, we know that the vector of ones **1** provides the minimal, but trivial solution $R = 0$, corresponding to a partition with all the nodes in a single group.

Non-trivial solutions for the cut minimisation therefore only emerge when we impose additional constraints on the minimisation of Eq. (3.8), for example, in terms of the size of the two clusters or by demanding that the indicator vector $\boldsymbol{s}$ is orthogonal to the vectors of ones: $\boldsymbol{s} \perp \mathbf{1}$. However, minimising Eq. (3.8) under the integer value constraints on s_i (and the additional constraints just discussed) is a difficult combinatorial optimisation problem in general.

If we neglect the integer constraints on s_i temporarily, but keep the constraint $\boldsymbol{s} \perp \mathbf{1}$, then it is straightforward to show (by decomposing $\boldsymbol{s}$ in the basis of eigenvectors of the Laplacian) that the vector with the smallest contribution to the cut size is the second eigenvector $\boldsymbol{v}_2$ of the Laplacian, which is associated to the smallest non-zero eigenvalue of the Laplacian. This eigenvalue is often called the spectral gap.[8] Unfortunately, setting $\boldsymbol{s}$ proportional to $\boldsymbol{v}_2$ is generally not an eligible solution, as the index vector $\boldsymbol{s}$ is supposed to contain only ± 1 entries. Nonetheless, such a spectral relaxation, in which we ignore the integer constraints on s_i, gives rise to many popular heuristics to minimise the cut size R subject to the constraint $\boldsymbol{s} \perp \mathbf{1}$. One particular heuristic, dating back to Fiedler (1973), is to choose an indicator vector $\boldsymbol{s}$ that is close to the second eigenvector $\boldsymbol{v}_2$. For this reason the vector $\boldsymbol{v}_2$ is often called the Fiedler eigenvector of the Laplacian. Since then, this line of reasoning has been applied to a broad class of optimisation problems, forming a popular class of heuristics often referred to as spectral methods. For instance, spectral methods have been proposed to optimise the so-called normalised cut (Shi & Malik, 1997) and the Newman–Girvan modularity (Newman, 2013), where eigenvectors of the normalised Laplacian and the so-called modularity matrix appear in place of the eigenvectors of the combinatorial Laplacian.

[8] Note that the term spectral gap is also used in the literature to refer to the difference of the dominant and the second dominant eigenvalue. As for the Laplacian, the relevant eigenvalues for our purposes are 0 and λ_2; this corresponds precisely to the notion of spectral gap used here.

3.3 Communities Defined by Node Equivalences: Disassortative Communities and Block Structures

As discussed in Section 3.1, community detection often looks for groups of nodes that are densely connected with each other. Finding groups of nodes such that the number of edges between nodes in different groups is as small as possible, for instance by minimising the cut, is a second, related but different perspective (Section 3.2). These two previous formalisations of community structure typically look for communities with more internal edges compared to some reference number, less edges between communities, or a combination of these two criteria. Although minimising a (normalised) cut size and maximising the internal number of links are closely related, there are important differences pertaining to the typical constraints and search space associated with these objective functions. For instance, to prevent finding trivial solutions that simply cut a single edge to disconnect the graph, we need to normalise the cut size in some way or specify the number of groups that we want to find. This information about the number of groups is typically not provided (or only implicitly, in terms of the null model) when using assortative community detection methods based on modularity maximisation and related principles.

However, there are also partitions that are not compatible with either of the two above-mentioned views that reveal commonalities in connectivity patterns between nodes. For instance, consider bipartite networks (i.e., networks composed of two groups of nodes such that all the edges are between these groups). The partition into these two groups is clearly not compatible with the concept of a community under either of the two perspectives we discussed so far: it would have a very low score for the modularity quality function for instance, and the cut between the two node groups would be maximal rather than minimal. Yet this partition identifies two different types of nodes in the network, whose existence may, for example, also be important for the dynamical behaviour of the network. Finding groups forming such disassortative communities, as they are called, is straightforward in the case of bipartite networks; for instance, by starting from a seed node and searching all the nodes at an even distance from that node. However, the problem becomes much more challenging when the network is not exactly bipartite, but almost bipartite,[9] or when other types of relationships between groups are present.

[9] Similarly, community detection is trivial when a graph is disconnected, and becomes challenging and interesting when the graph is *almost disconnected*.

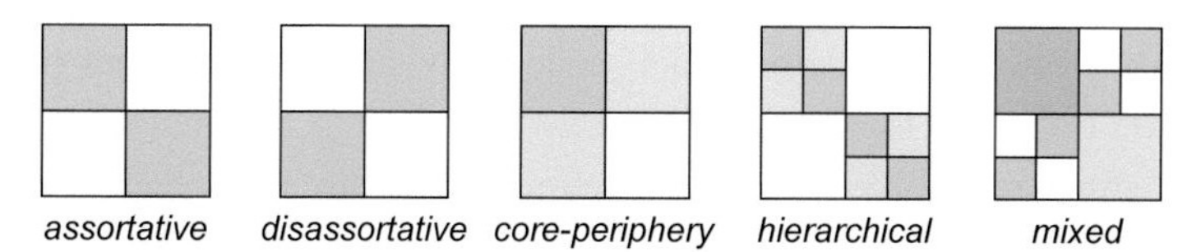

Figure 4 Affinity matrices for different block structures. Different types of block structures can be observed in networks, including assortative communities, where most of the connections are concentrated inside communities, the opposite notion of disassortative communities, core–periphery structure, where connections are mostly present inside a core and between this core and the network periphery, and hierarchical or other mixed structures.

Detecting such disassortative communities is related to the notions of node equivalence, role extraction, and block modelling (Wasserman & Faust, 1994). At the heart of these concepts, there is the assumption that different types of nodes exist and that these node types, or node roles, can be identified by their connectivity patterns. Associating one meta-node for each group of similar nodes leads to a simplified representation of the network, called the reduced graph or the image graph. The same information can also be visualised by reordering the nodes based on their group membership, leading to an adjacency matrix with a block structure (see Figure 4). Assortative communities correspond to a particular class of role assignments, where for each role nodes mainly interact with nodes in the same role, thus leading to a block diagonal structure for the adjacency matrix after an appropriate node permutation. However, many other kinds of role interactions may be defined such as core–periphery (Rombach et al., 2014) or block cycle models for food webs (Van Lierde, Chow, & Delvenne, 2019).

Role extraction typically relies on the definition of an equivalence between nodes. The first measure of node equivalence, proposed in Lorrain and White (1971), has been structural equivalence, which posits that two nodes are equivalent if they have exactly the same neighbours. Note that in the case of bipartite networks, this implies that in the adjacency matrix the lines and columns of two equivalent nodes are equal. Unfortunately, this measure of equivalence is very restrictive and leads to the extraction of many small roles in real-world networks. This observation led to the development of alternative measures built on the same principle. For instance, two nodes are considered regularly equivalent (Everett & Borgatti, 1994) if they are connected to the nodes in the same equivalence classes, independently of the number of such connections. Regular equivalence is clearly a relaxation of structural equivalence, as structural equivalence implies regular equivalence, but the opposite is not true (Brandes, 2005, Chapter 9).

Nonetheless, most of these (exact) node equivalence formulations proved to be too restrictive for dealing with real-world data. A probabilistic relaxation of structural equivalence was therefore constructed by Holland, Laskey, and Leinhardt (1983) via the so-called stochastic block model (SBM). Given n nodes divided into C groups, a standard SBM (Abbe, 2017) is defined by a $C \times C$ affinity matrix $\mathbf{\Omega}$ and a partition $\mathcal{P} = \{\mathcal{A}_1, \mathcal{A}_2, \ldots, \mathcal{A}_C\}$ of the nodes into communities. The SBM then posits that a link between two nodes i, j belonging to classes $\mathcal{A}_\alpha$ and $\mathcal{A}_\beta$ is described by a Bernoulli random variable with probability:

$$p_{ij} := \mathbb{P}(A_{ij}) = \Omega_{\alpha\beta}. \tag{3.9}$$

The SBM thus assumes that there exists a latent probabilistic process that generated the network that is dependent only on the group labels.

Note that the affinity matrix encodes the block structure of the graph (see Figure 4), and may be seen as a weighted network of groups. Under the SBM, nodes within the same equivalence class thus have exactly the same probabilities to connect to nodes of another class. Accordingly, nodes within one group are said to be *stochastically equivalent*. This is precisely the same condition as structural equivalence, but rather than formulating the node equivalence in term of the adjacency matrix of the graph, stochastic equivalence is concerned with the expected adjacency matrix induced by the (latent) node connection probabilities (see also Figure 9 and the associated discussion in Section 5.3).

An SBM can be seen as a generalisation of the classical Erdős–Rényi model, in which case the nodes would all belong to a single group connected with probability q. Instead, the SBM allows links for each combination of the group labels to have a different link probability. Finding the latent groups of nodes in a real-world network now amounts to inferring the model parameters (the matrix $\mathbf{\Omega}$ and the partition $\mathcal{P}$) of the SBM that provide the best fit for the observed network, for example, provide the SBM with the highest likelihood. Because inferring the most likely SBM typically results in grouping nodes based on their degree in empirical networks with broad degree distributions, it is advantageous to include a degree correction into the model. This leads to the degree-corrected SBM (DCSBM) (Dasgupta, Hopcroft, & McSherry, 2004; Karrer & Newman, 2011), in which the probability for a link to appear between two nodes i, j is assumed to be of the form

$$p_{ij} \sim k_i k_j \Omega_{\alpha\beta}, \tag{3.10}$$

where α and β are the labels of the communities nodes i and j belong to. Observe that in addition to the dependence on the class labels α, β, the probability p_{ij} is now influenced by the degrees k_i, k_j of the respective nodes, similar to the soft configuration model.

In contrast with the community detection methods that we have seen so far, maximising the likelihood of an SBM does not necessarily aim to maximise some internal density, or alternatively, to minimise a cut. Instead we assume that the data has been generated according to a given model and look for the model parameters that would most likely have generated the network data that we observe. This network may contain assortative or disassortative communities, or a mixture of the two. For instance, for a bipartite network, fitting a (degree-corrected) stochastic block model with two groups should find the bipartite split. It can also happen that competing maxima of the likelihood are associated to different types of block structures (Peel, Larremore, & Clauset, 2017).

We remark that the number of groups has to be specified to fit a (DC)SBM to any observed network data. To make this approach operational for community detection when the number of groups is not known a priori, one thus has to employ some kind of model selection mechanism to choose an appropriate number of groups. One powerful approach for model selection is to employ a Bayesian procedure and select the number of groups based on a minimum description length principle (Peixoto, 2019).

Ensemble of Graphs versus Single Realisation

A conceptual difference between the SBM perspective and the perspective employed, for example, when minimising a cut, is that the former aims to provide a generative model for the network and considers the observed network to be but one observation from a possible ensemble of networks (Newman et al., 2003), whereas the latter considers the observed network as a fixed entity. In order to illustrate the different answers provided by each type of method, let us consider a real-world graph generated by a possibly complex random assignment of edges. From a cut minimisation perspective, given that network, we would like to find groups of weakly connected nodes. For instance, we may want to help to stop the spread of a rumour and thus partition the network into weakly connected modules of nodes. Whether or not the modular structure that is observed resulted from random fluctuation in the creation of the graph, these modules will be relevant for our task. In other words, we consider the single observed network independently of the mechanisms that may have generated it.

The problem would appear completely different from an SBM perspective. Let us assume for simplicity of our argument that the observed graph has likely been generated according to a realisation of an Erdős–Rényi graph.[a] In this case, a model selection approach paired with an SBM is expected to find that that the ER model with no communities is sufficient to explain the data, as the variations in the data can already be explained by random fluctuations rather than by hidden group labels.

This example illustrates that different motivations for community detection may find different useful answers even for the very same network. Rather than looking at community detection as a generic tool that is supposed to work in a generic context, considering the application and the modelling questions in mind is thus critical when choosing between or comparing different methods, especially if they are based on different principles. This statement should be kept in mind for Section 6, where we will develop flow-based, also called dynamical, community detection methods instead of the structural, combinatorial methods presented so far.

[a] Note that ER networks in a sparse regime are in fact often disconnected.

Before closing, let us note that there are certain connections between the Newman–Girvan modularity of Section 3.1 and the notion of block modelling. First, it can be shown that when modularity is equipped with a 'resolution parameter' its optimisation leads to the same partition as that of the so-called planted partition model with an appropriate, fixed number of communities (Newman, 2016). The planted partition model is a simplification of stochastic block models where the affinity matrix $\boldsymbol{\Omega}$ can only take two values, p_{in} and p_{out}, defining the probability for an edge inside or across communities. In addition, starting from the formulation (3.3) of modularity as the trace of the $C \times C$ matrix

$$\boldsymbol{Q} = \frac{1}{2m}\left[\boldsymbol{H}^T\left[\boldsymbol{A} - \frac{\boldsymbol{k}\boldsymbol{k}^\top}{2m}\right]\boldsymbol{H}\right], \tag{3.11}$$

one can search for other types of block structures by characterising the goodness of a partition with a subset of the elements of $\boldsymbol{Q}$, not necessarily on its diagonal. This can be done, for instance, by specifying a block structure with a binary affinity matrix $\Omega_{\alpha\beta} = 1$ for connected groups, and zero otherwise (Reichardt & White, 2007), and optimising

$$\sum_{\alpha,\beta=1}^{c} \mathcal{Q}_{\alpha\beta}\Omega_{\alpha\beta}. \tag{3.12}$$

That method allows one to find the best partition with a given number of groups and a given block structure. Within this framework, finding disassortative communities can be done by maximising the off-diagonal elements of (3.11), which is equivalent to minimising the standard Newman–Girvan modularity.

3.4 Further Discussion and References

Community detection has been an active field of research for the last 15 years, with contributions and applications from and to different disciplines. In this section, we have only scratched the surface of the question, focusing on methods that will serve as a basis for the discussions that will follow, especially in Sections 6 and 7. For a more thorough and complete overview of community detection, we refer the reader to the review papers (Fortunato, 2010; Fortunato & Hric, 2016; Porter, Onnela, & Mucha, 2009; Schaeffer, 2007; Schaub et al., 2017). Another insightful resource is the recently edited book (Doreian et al., 2020), whose collection of chapters provides introductory and more advanced topics on clustering and block modelling in networks. Specialised aspects of community detection are also discussed in more focused references, for instance Malliaros and Vazirgiannis (2013) for directed networks and Rossetti and Cazabet (2018) for temporal networks.

4 Timescale Separation and Dynamics on Modular Networks

In this section and the next one, we discuss timescale separation and symmetries, two concepts often used to simplify the description of a dynamical system. In particular, we explain how these concepts are linked with modular network structure in case of a linear dynamics. Our exposition builds on Schaub et al. (2019b) but provides a substantial amount of additional detail.

Timescale separation is the phenomenon when certain state variables of a dynamical system evolve much faster than other state variables. Accordingly, we may group the state variables into fast variables and slow variables and attempt a simplified analysis. If we are interested in the system behaviour over short time horizons, we may treat the slow state variables as approximately constant and analyse solely the fast variables; if we are interested in the system behaviour over long time horizons, we may treat the fast variables as if they are negligible (e.g., we may consider that they have equilibrated and are not evolving any more). In general, this analysis of the system behaviour is only approximate. However, the error made can be bounded under certain assumptions. If this is the case, we obtain a simplified system description that may be significantly less complex than the original system. Generally, there may be multiple timescales present in the system rather than just two, and we may be able to divide our state variables into multiple groups, leading to a simplified multiscale description.

How can the separation of timescales be related to networks with modular structure? In the following, we will consider general linear dynamics on modular networks and will see that timescale separation is in this case closely related to a separation of eigenvalues in the spectrum of the system matrix (the graph) governing the state evolution of the system.

4.1 Timescale Separation for General Dynamics

Let us exemplify the concept of timescale separation more formally with the following example of a two-dimensional dynamical system:

$$\frac{dx}{dt} = f(x,y), \tag{4.1a}$$

$$\epsilon^{-1}\frac{dy}{dt} = g(x,y). \tag{4.1b}$$

We assume for simplicity that f, g are bounded functions of order $\mathcal{O}(1)$, and $\epsilon \ll 1$ is a small constant relative to those bounds. Observe that in the above system the state variable $x(t)$ changes much more rapidly than $y(t)$, since $dy/dt = \epsilon g(x,y)$ is small by construction. Alternatively we may define the slow time variable $\tau := \epsilon\, t$, such that Eq. (4.1b) can be rewritten as $dy/d\tau = g(x,y)$.

This rewriting emphasizes the *separation of timescales* in the dynamics: y evolves according to the slow timescale τ, whereas x evolves according to the faster timescale t.

As alluded to at the beginning of this section, when a timescale separation is present in a system, the dynamics of x and y are approximately decoupled in two different regimes: for the fast behaviour, we may simply concentrate on x and assume y to be approximately constant; for the slow, long-term behaviour we may focus on y and assume that x is in its asymptotic state for the given value of y, thus forgetting about the detailed dynamics of x. When several *distinct* timescales are present, we can similarly approximate the dynamics over particular timescales by reduced dynamics that can be obtained by finding quasi-invariant subspaces in the original system. These concepts emerge naturally in the study of networked dynamics, as we discuss below.

4.2 Timescale Separation for Linear Network Dynamics

Let us now consider timescale separations in linear dynamical systems defined on a network. To do so, we continue our analysis of the system $\dot{\boldsymbol{x}} = \boldsymbol{F}\boldsymbol{x}$ with $\boldsymbol{x}(0) = \boldsymbol{x}_0$ of the previous sections. We assume that the system matrix $\boldsymbol{F}$ is at least marginally stable, which implies that the dynamics remains bounded for all times. To reveal the characteristic timescales of the process, it is useful to start from the spectral expansion (2.11) of the solution

$$\boldsymbol{x}(t) = \sum_{i=1}^{n} e^{\lambda_i t}\, \boldsymbol{v}_i \boldsymbol{u}_i^\top \boldsymbol{x}_0 = \sum_{i=1}^{n} \left(\boldsymbol{u}_i^\top \boldsymbol{x}_0\right) e^{\lambda_i t}\, \boldsymbol{v}_i, \tag{4.2}$$

showing that the timescales of the process are dictated by the eigenvalues of the matrix $\boldsymbol{F}$. Each eigenmode (right eigenvector) decays with a characteristic timescale $\tau_i = -1/\lambda_i$. Hence, if there are large differences (gaps) between eigenvalues, the system will have timescale separation. For instance, if the k largest (dominant) eigenvalues $\{\lambda_1, \ldots, \lambda_k\}$ are well separated from the remaining eigenvalues such that $\lambda_k \gg \lambda_{k+1}$, the eigenmodes associated with $\{\lambda_{k+1}, \ldots, \lambda_n\}$ become negligible for $t > -1/\lambda_{k+1}$ and it follows that the system can be effectively described by the k dominant eigenmodes for $t > -1/\lambda_{k+1}$. More technically, we say that the first k eigenvectors form a dominant invariant subspace of the dynamics and there exists an associated lower-dimensional ($k < n$) approximate description of the dynamics on the network after the timescale $\tau \approx -1/\lambda_{k+1}$.

To see this explicitly, we define the approximate system dynamics $\hat{\boldsymbol{x}}_{\boldsymbol{k}}(t) = \sum_{i=1}^{k} \left(\boldsymbol{u}_i^\top \boldsymbol{x}_0\right) e^{\lambda_i t}\, \boldsymbol{v}_i$ that considers only the first k dominant eigenmodes of the system and compute how the approximation error $\epsilon_k(t) = \|\boldsymbol{x}(t) - \hat{\boldsymbol{x}}_{\boldsymbol{k}}(t)\|$ evolves over time:

$$\epsilon_k(t) = \|\boldsymbol{x}(t) - \hat{\boldsymbol{x}}_{\boldsymbol{k}}(t)\| = \left\| \sum_{i=k+1}^{n} \left(\boldsymbol{u}_i^\top \boldsymbol{x}_0\right) e^{\lambda_i t}\, \boldsymbol{v}_i \right\| \tag{4.3a}$$

$$\leq \sum_{i=k+1}^{n} e^{\lambda_i t} \left\| \left(\boldsymbol{u}_i^\top \boldsymbol{x}_0\right) \boldsymbol{v}_i \right\| \leq e^{\lambda_{k+1} t} \sum_{i=k+1}^{n} \left\| \left(\boldsymbol{u}_i^\top \boldsymbol{x}_0\right) \boldsymbol{v}_i \right\| \tag{4.3b}$$

$$\leq e^{\lambda_{k+1} t} \epsilon_k(0). \tag{4.3c}$$

As the above computations show, the initial error decays exponentially with a rate of at least λ_{k+1} (e.g., after $t \approx -3/\lambda_{k+1}$ the initial error of neglecting all but the first k modes has been reduced by a factor of $e^{-3} \approx 0.05$). In contrast, while the amplitude of the remaining signal $\hat{\boldsymbol{x}}_{\boldsymbol{k}}(t)$ will have decreased as well (due to the system stability), this decrease will be at a rate of at most λ_k, which is far slower due to the separation of timescales. Indeed, for a marginally stable system, some of the first k modes might not have decayed at all and can thus not be neglected. Similar arguments can be made when focusing only on the fast timescale or dealing with multiple separated timescales. This line of reasoning is at the core of model order reduction methods for dynamical systems.

As we have seen, the spectral properties of the coupling matrix $\boldsymbol{F}$ of the network dynamics are responsible for the timescale separation. To connect this finding with modular structure, we need to understand how this structure can influence the spectral properties of the linear operator $\boldsymbol{F}$. As we will see, for a large class of stable dynamics, including diffusion processes as well as consensus and opinion formation models, a timescale separation can be induced by *localised substructures* in the graph.

4.3 Assortative Modular Network Structure and Timescale Separation

In this section, we will discuss how assortative modular structure can give rise to a separation of timescales in a network. We will first discuss a setup in which the network is arbitrary but fixed, and we can model the network as a perturbation of a perfectly assortative modular network with C disconnected components. Thereafter, we will discuss how this picture can be translated into a stochastic setup, in which there is a random generative process from which the network is drawn.

4.3.1 Matrix Perturbation Theory for Assortative Modular Networks

For simplicity, let us start with the concrete example of a consensus dynamics $\dot{\boldsymbol{x}} = -\boldsymbol{L}\boldsymbol{x}$ with initial condition $\boldsymbol{x}(0) = \boldsymbol{x}_0$, which takes place on a network composed of C modules with an adjacency matrix of the form

$$\boldsymbol{A} = \begin{pmatrix} \boldsymbol{A}_1 & & & \\ & \boldsymbol{A}_2 & & \\ & & \ddots & \\ & & & \boldsymbol{A}_C \end{pmatrix} + \boldsymbol{A}_{\text{noise}} =: \boldsymbol{A}_{\text{structure}} + \boldsymbol{A}_{\text{noise}}. \tag{4.4}$$

Each block $\boldsymbol{A}_\alpha$ in the block-diagonal matrix $\boldsymbol{A}_{\text{structure}}$ is supposed to correspond to a densely connected graph, and $\boldsymbol{A}_{\text{noise}}$ is a weak perturbation of this strong assortative modular structure. Let us denote the dimension of each block by n_α, such that $\sum_\alpha n_\alpha = n$.

To see how a structure like the one above can give rise to a separation of timescales in the consensus dynamics, we have to assess what the spectrum of the corresponding Laplacian $\boldsymbol{L} = \boldsymbol{L}_{\text{structure}} + \boldsymbol{L}_{\text{noise}}$ looks like. To this end, we treat $\boldsymbol{L}_{\text{noise}}$ as a perturbation of $\boldsymbol{L}_{\text{structure}}$ and employ Weyl's perturbation theorem to obtain bounds for the eigenvalues of $\boldsymbol{L}$. Hence, we first consider the case where $\boldsymbol{L}_{\text{noise}} = \mathbf{0}$ (i.e., the graph consists of C disconnected components). It follows from the properties of Laplacian matrices that in this case $\boldsymbol{L}$ has an eigenvalue $\lambda = 0$ with multiplicity C. In fact, for $\boldsymbol{L}_{\text{noise}} = \mathbf{0}$ the consensus dynamics completely decouples and the eigenspace associated to the zero eigenvalues can be spanned by C indicator vectors $\boldsymbol{h}^{(1)}, \ldots, \boldsymbol{h}^{(C)}$ of the connected components of the network, corresponding to the C diagonal blocks in the adjacency matrix $\boldsymbol{A}$:

$$[\boldsymbol{h}^{(\alpha)}]_j = \begin{cases} 1 & \text{if } 1 + \sum_{\beta<\alpha} n_\beta \le j \le \sum_{\beta\le\alpha} n_\beta \\ 0 & \text{otherwise.} \end{cases} \tag{4.5}$$

In this extreme case of an assortative community structure, we thus have at least one clear separation of timescales: the zero eigenvalues correspond to modes with no time evolution at all, whereas all other eigenmodes will be associated with an exponentially decaying signal with rate $\lambda_i(\boldsymbol{L}) > 0$ for $i > C$.

Let us now consider the case where $\boldsymbol{A}_{\text{noise}} \neq \mathbf{0}$ but may be considered as a small perturbation of $\boldsymbol{A}_{\text{structure}}$. In this case we can use the so-called Weyl's perturbation theorem (Bhatia, 2013, Chapter III.2) to gain insight about the spectrum of $\boldsymbol{L}$:

Theorem 1 (Weyl's perturbation theorem) *Let $\boldsymbol{M}$, $\boldsymbol{P}$ be $n \times n$ Hermitian matrices.*[10] *Let $\boldsymbol{\lambda}^{\downarrow}(\boldsymbol{M})$ denote the vector of decreasingly ordered eigenvalues of $\boldsymbol{M}$, such that $\lambda_1^{\downarrow}(\boldsymbol{M}) \geq \ldots \geq \lambda_n^{\downarrow}(\boldsymbol{M})$, and define $\boldsymbol{\lambda}^{\downarrow}(\boldsymbol{P})$ analogously. Then for each $j = 1, \ldots, n$*

$$|\lambda_j^{\downarrow}(\boldsymbol{P}) - \lambda_n^{\downarrow}(\boldsymbol{M})| \le \max_{\ell} |\lambda_\ell^{\downarrow}(\boldsymbol{P}) - \lambda_n^{\downarrow}(\boldsymbol{M})| \le \|\boldsymbol{P} - \boldsymbol{M}\|. \tag{4.6}$$

Described in words, Weyl's inequalities state that the ordered eigenvalues of the perturbed matrix $\boldsymbol{P}$ are close to the eigenvalues of the unperturbed matrix $\boldsymbol{M}$ (i.e., their absolute difference is at most given by the spectral norm $\|\boldsymbol{P} - \boldsymbol{M}\|$, where the spectral norm of a matrix is defined as its largest singular value).

We can apply Weyl's perturbation theorem to our specific example by setting $\boldsymbol{P} = \boldsymbol{L}$ and $\boldsymbol{M} = \boldsymbol{L}_{\text{structure}}$, which means that the bound $\|\boldsymbol{P} - \boldsymbol{M}\| = \|\boldsymbol{L}_{\text{noise}}\|$ is precisely the spectral norm of the perturbation. Hence, if there was a significant eigenvalue gap in $\boldsymbol{L}_{\text{structure}}$ and the perturbation $\boldsymbol{L}_{\text{noise}}$ is sufficiently small in spectral norm, there will still be an eigenvalue gap in $\boldsymbol{L}$. This means that the

[10] When a matrix has only real entries, as considered throughout this Element, it is Hermitian if and only if it is symmetric.

consensus dynamics on our modular network will also display a separation of timescales. An illustration of this is given in Figure 5.

Note that Weyl's perturbation theorem is not specific to the consensus dynamics discussed above, but can be applied to any Hermitian matrix (or symmetric matrix, for real-valued matrices). In particular, let us consider again a dynamics on an undirected network governed by a linear operator of the form $\boldsymbol{F} = \boldsymbol{D}\boldsymbol{A}_{\mathcal{G}}\boldsymbol{D}^{-1}$, as discussed above. Observe that $\boldsymbol{F}$ can be related via a similarity transform to the symmetric matrix $\boldsymbol{F}_{\text{sym}} = \boldsymbol{D}^{-1/2}\boldsymbol{A}_{\mathcal{G}}\boldsymbol{D}^{-1/2}$ and both matrices thus have the same spectrum. Therefore we can, with little extra work, employ Weyl's theorem for any dynamics on an undirected network, if we can express the matrix $\boldsymbol{F}_{\text{sym}}$ as a linear combination of a 'structure' part associated to slow eigenvalues and a 'noise' part with a comparably small spectral norm. Importantly, if the network under consideration has indeed an assortative modular structure, then we can typically express $\boldsymbol{F}_{\text{sym}}$ as a low-rank matrix describing the modules that get perturbed by a small (sparse) noise component. This is similar to the composition of the consensus dynamics considered above and highlights how assortative modules can more generally give rise to a timescale separation in a linear system dynamics.

4.3.2 Stochastic Assortative Modular Structure and Separation of Timescales

Thus far we have considered undirected networks whose adjacency matrix was arbitrary but fixed, and associated the presence of a timescale separation to the presence of a low-rank component with slow eigenmodes that is perturbed by noise. Let us now consider the case in which the observed network has been drawn from a random graph model with an assortative modular structure.

For instance, the adjacency matrix might have been generated from a planted partition model, a simplified assortative variant of the more general stochastic block model (SBM) mentioned in Section 3.3. Recall that just like the SBM, the planted partition model posits that each node is assigned a group label. Then, nodes with the same group labels are connected with probability p_{in}, and nodes with different group labels are connected with probability p_{out}. We will denote the indicator matrix of the partition associated to this model by $\boldsymbol{H}$ where, as before, $H_{i\alpha} = 1$ if node i belongs to community $\mathcal{A}_\alpha$ and $H_{i\alpha} = 0$ otherwise.

Consider now a linear dynamics evolving on an undirected graph with adjacency matrix $\boldsymbol{A}_{\text{PP}}$ drawn from the planted partition. For simplicity, let us focus on a normalised Laplacian dynamics evolving according to $\dot{\boldsymbol{x}} = -\boldsymbol{\mathcal{L}}\boldsymbol{x}$, though similar arguments can be applied for other system matrices $\boldsymbol{F}$. To show that a structured random graph model such as the planted partition model does indeed induce a separation of timescales if there is well-defined community structure

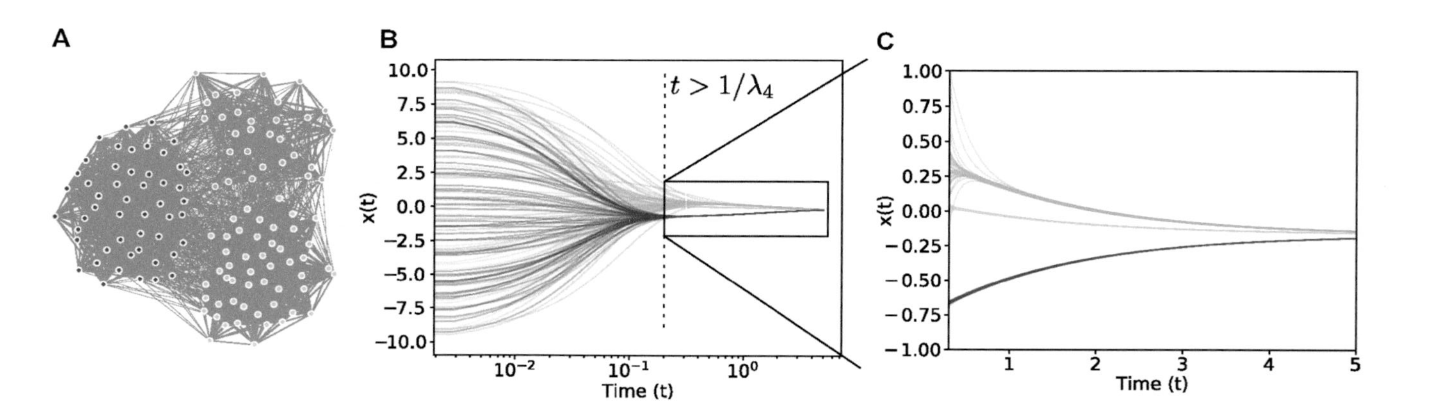

Figure 5 Consensus dynamics on a structured network. A Visualisation of a network with three groups and an adjacency matrix of the form (4.4). **B** When observing a consensus dynamics on this network, there is a clear timescale separation: after $t \approx 1/\lambda_4 = 0.2$, approximate consensus is reached within each group (indicated by color). Eventually global consensus is reached across the network.

in terms of the model parameters, we will again leverage Weyl's eigenvalue perturbation theorem (Theorem 1).

To invoke Weyl's perturbation theorem, we again have to find a suitable decomposition of the normalised Laplacian $\boldsymbol{\mathcal{L}}$ into a 'structure' component and a 'noise' component perturbing this structure. Unlike in the deterministic setting discussed above, now the graph is random and, accordingly, the normalised Laplacian is a random matrix, so our decomposition will need to involve some random matrices as well. In the case of the normalised Laplacian it is useful to consider the Laplacian $\boldsymbol{\mathcal{L}}(\mathbb{E}[\boldsymbol{A}])$ of the expected adjacency matrix under the planted partition model as the structure component, and the deviation $\boldsymbol{\mathcal{L}}_{\text{noise}} = \boldsymbol{\mathcal{L}}(\boldsymbol{A}) - \boldsymbol{\mathcal{L}}(\mathbb{E}[\boldsymbol{A}])$ of the normalised Laplacian of the observed adjacency matrix from the normalised Laplacian of the expected adjacency matrix as our noise component:

$$\boldsymbol{\mathcal{L}} = \boldsymbol{\mathcal{L}}(\mathbb{E}[\boldsymbol{A}]) + \boldsymbol{\mathcal{L}}(\boldsymbol{A}) - \boldsymbol{\mathcal{L}}(\mathbb{E}[\boldsymbol{A}]) = \boldsymbol{\mathcal{L}}_{\text{structure}} + \boldsymbol{\mathcal{L}}_{\text{noise}}. \tag{4.7}$$

Observe that $\boldsymbol{\mathcal{L}}(\mathbb{E}[\boldsymbol{A}])$ in the above decomposition is in fact a low-rank matrix, shifted by an identity matrix. Specifically, $\boldsymbol{\mathcal{L}}(\mathbb{E}[\boldsymbol{A}])$ can be written as $\boldsymbol{\mathcal{L}} = \boldsymbol{I} - \boldsymbol{H}\boldsymbol{\Theta}\boldsymbol{H}^\top$ for some matrix $\boldsymbol{\Theta} \in \mathbb{R}^{C\times C}$ that will depend on the parameters of the planted partition model. For an assortative planted partition model, $\boldsymbol{\mathcal{L}}(\mathbb{E}[\boldsymbol{A}])$ will thus have at most C small eigenvalues corresponding to slow timescales, whereas the remaining eigenvalues will be equal to 1. To invoke Weyl's perturbation theorem, we thus need to guarantee that $\|\boldsymbol{\mathcal{L}}_{\text{noise}}\| = \|\boldsymbol{\mathcal{L}}(\boldsymbol{A}) - \boldsymbol{\mathcal{L}}(\mathbb{E}[\boldsymbol{A}])\|$ is small in a suitable sense (i.e., that the normalised Laplacian of the observed adjacency matrix is close to the normalised Laplacian of the expected adjacency matrix). However, $\boldsymbol{A}$ is a random matrix, and we thus cannot guarantee that the norm $\|\boldsymbol{\mathcal{L}}_{\text{noise}}\|$ will be small in general. For instance, there is a small but nonzero probability that our adjacency matrix will be almost empty, in which case we will have a large perturbation $\|\boldsymbol{\mathcal{L}}_{\text{noise}}\|$ that can be close to the largest eigenvalue $\lambda_1^{\downarrow}(\boldsymbol{\mathcal{L}}(\mathbb{E}[\boldsymbol{A}]))$ of the normalised Laplacian of the expected adjacency matrix. Hence, the best we can hope for is to guarantee that $\|\boldsymbol{\mathcal{L}}_{\text{noise}}\|$ will be small with a high probability. The typical bound one aims to obtain in this scenario is thus to have the norm of the noise term $\|\boldsymbol{\mathcal{L}}_{\text{noise}}\|$ to be smaller than a certain number δ with a (high) probability $1 - \epsilon(\delta)$ that may depend on the chosen bound δ. Such a result can indeed be proven using so-called concentration inequalities,[11] which bound the difference of a random variable (matrix) from its expectation (Wainwright, 2019).

[11] Note that a probabilistic bound on $\|\boldsymbol{\mathcal{L}}_{\text{noise}}\|$ is not a concentration inequality itself, since the expected value of the normalised Laplacian $\mathbb{E}[\boldsymbol{\mathcal{L}}(\boldsymbol{A})]$ is not equal to $\boldsymbol{\mathcal{L}}(\mathbb{E}[\boldsymbol{A}])$.

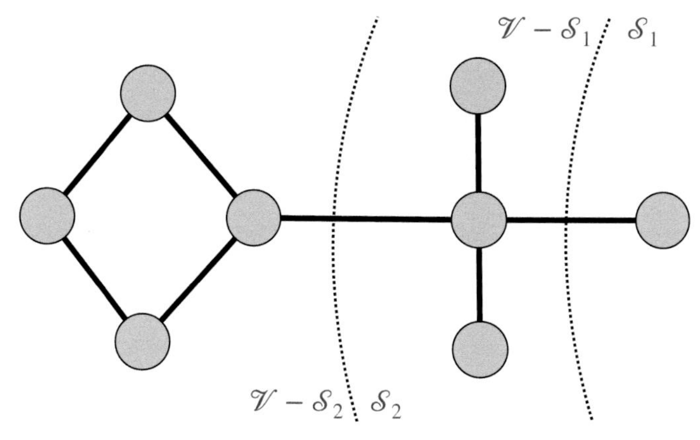

Figure 6 Conductance of a graph. The conductance of the sets of nodes $\mathcal{S}_1$ and $\mathcal{S}_2$, defined by Eq. (4.8), are equal to 1 and 1/7 respectively. Both sets have the same cut size but different volumes. The conductance of $\mathcal{S}_2$ is smaller because it provides a more balanced division of the graph.

4.4 Non-assortative Network Structures and Directed Networks

4.4.1 Beyond Assortative Network Sructures

In the preceding discussion, we have focused on assortative community structure, that is, a notion of community that posits that a community should have a large number of connections to nodes within its own group and only relatively few links to nodes outside. However, a separation of timescales may also be induced by network structures that are *not* block-homogeneous in this sense.

For instance, for many networks embedded in space, such as power grids, road networks, or other supply and infrastructure networks, there can be well-defined network modules that are only weakly coupled to the rest of the network. Such modules may be detected when focusing on a cut-based notion of modular structure (i.e., the number of links between communities should be minimal). Yet, in contrast to typical assortative community structure, these modules may internally not be very densely connected, for example, because of the geometry of the space in which the network is embedded or some other constraints (such as connection costs). As a concrete example, think of a street network inside a city that is divided by a river which poses a natural barrier to the connectivity of the network and divides the city into two parts. In many circumstances, it will be meaningful to think of such a network as split into two modules. Clearly, such a split can influence a dynamical process on a network, too.

Indeed, presuming we have a diffusion dynamics governed by a normalised Laplacian $\boldsymbol{\mathcal{L}}$, the presence of a sparse cut in the connectivity can be related to the spectral properties of the Laplacian via the Cheeger inequality and hence can lead to a separation of timescales. To state the Cheeger inequality properly, we define for every set of nodes $\mathcal{S} \subset \mathcal{V}$ the conductance $\phi_{\mathcal{S}}$ as

$$\phi_{\mathcal{S}} = \left\{ \frac{\sum_{i \in \mathcal{S}, j \notin \mathcal{S}} A_{ij}}{\min\{\mathrm{vol}(\mathcal{S}), \mathrm{vol}(\mathcal{V} - \mathcal{S})\}} \right\}, \tag{4.8}$$

where $\mathrm{vol}(\mathcal{S}) := \sum_{i \in \mathcal{S}} k_i$ is the total connectivity of the set, called the volume of $\mathcal{S}$. The conductance of a graph is then defined as the minimal conductance for all possible node sets: $\phi_{\mathcal{G}} = \min_{\mathcal{S}} \phi_{\mathcal{S}}$. Note that the graph conductance is small if there exist two sets of nodes that are of similar size and have few connections between them. We can now state the Cheeger inequality, which relates the graph conductance to the second smallest eigenvalue of the normalised Laplacian as follows:[12]

$$\frac{\phi_{\mathcal{G}}^2}{2} < \lambda_2 \leq 2\phi_{\mathcal{G}}\,. \tag{4.9}$$

The Cheeger inequality shows that a small value of the graph conductance is associated to a small spectral gap λ_2 and hence a comparably slow relaxation of the diffusion dynamics to its stationary state. This means that if the network can be divided into two well-separated node sets with a small cut between them, this bottleneck will slow down diffusion and can thus lead to a separation of timescales. Note that this result holds irrespective of whether the nodes in these two sets are homogeneously connected with each other in a dense, assortative way, corresponding precisely to the situation outlined above. Indeed, many networks contain such natural substructures which are not assortative yet act effectively as a dynamical module over a particular timescale (Schaub et al., 2015; Schaub et al., 2012).

4.4.2 Timescale Separations on Directed Networks

While we focused on a linear dynamics on undirected networks above, a separation of timescales is, of course, also possible for a linear dynamics on a directed network. As alluded to in Section 2.3.3, analysing such directed networks is, however, mathematically far more complicated. While the timescales of a linear system on a general directed network will still be governed by the eigenvalues of the matrix, these eigenvalues will generally not be real anymore but complex. This implies, for instance, that the system can now display oscillatory dynamics which were not possible for the dynamics on the undirected graphs considered before. More importantly, from a theoretical point of view, the foundations of this section based on Weyl's perturbation theorem (Theorem 1) are not applicable any more.

[12] Note that there are several variations of the Cheeger inequality (e.g., for different matrices describing the graph).

Nonetheless, there are several studies that discuss the impact of network structure on the timescales present in a linear dynamical system taking place on a directed graph. In particular, for linear diffusion dynamics, it can be shown that networks with a block-cyclic structure (Van Lierde et al., 2019) exhibit a separation of timescales associated to slow oscillations between groups of nodes in a network. Such groups of nodes may accordingly be seen as communities within the network, which are here to be understood in a dynamical rather than a structural fashion. Similar ideas have also been put forward in Banisch and Conrad (2015) and Conrad, Weber, and Schütte (2016), where communities are defined as directed graph structures that retain probability flow over long timescales, and thus are associated to a separation of timescales in the network. The Markov stability framework, which we discuss in detail in Section 6, provides another example for exploiting this kind of timescale separation phenomena to define community structure in networks.

4.5 Further Discussion and References

The topics covered in this section range from fairly classic to relatively new results and methods. The concept of timescale separation is a well-established technique for the analysis of dynamical systems, and more extensive discussions can be found in most standard texts on dynamical systems. Of particular relevance to our context is the classic result by Simon and Ando (1961) on the aggregation of the states in Markov chains, which can be directly related to diffusion processes and random walks on graphs. Matrix perturbation theory is covered in a range of standard texts on (Numerical) Linear Algebra, including Golub and Van Loan (2013), Stewart (2001), and Bhatia (2013). Many results on the perturbation of eigenvalues can be traced back to Weyl. Our presentation of Weyl's eigenvalue perturbation theorem, in Theorem 1, is based on Bhatia (2013), which contains a far more extensive discussion on this and related results. The use of concentration inequalities to study (perturbations of) random matrices is a far more recent topic. Specifically in the context of analysing various high-dimensional problems in Machine Learning and Data Science, including inferring the partitions within stochastic block models (see, e.g., Lei & Rinaldo (2015); Rohe, Chatterjee, Yu et al. (2011)), these tools have gained prominence in the literature recently. For an introduction to these techniques see, for example, Wainwright (2019).

In this section, we always considered situations with non-overlapping communities, as the nodes naturally belonged to one community, and edges were the objects bridging between them. Interesting venues of future research include a study of how linear dynamics is affected by overlapping communities, where so-called broker nodes (Burt, 2004) belong to more than one community.

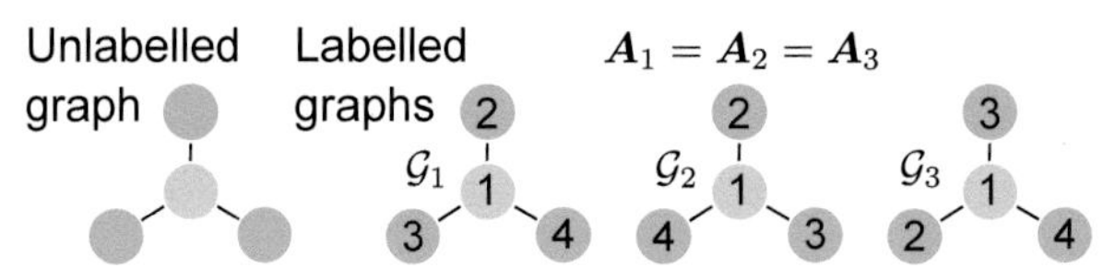

Figure 7 Graphs with symmetry. Left: an unlabelled star graph that is highly symmetric. Right: Assigning labels to the nodes as indicated leads to exactly the same adjacency matrix $\boldsymbol{A}_1 = \boldsymbol{A}_2 = \boldsymbol{A}_3$.

This question would be even more challenging in the situation of pervasively overlapping communities, where each node may belong to several communities, as in social networks where people usually belong to more than one social circle (Ahn et al., 2010).

5 Symmetries and Dynamics on Modular Networks

In the previous section, we introduced the concept of timescale separation as a means to establish a simplified description of a dynamical system on a network. In this section, we concentrate on how the presence of symmetries can provide another pathway towards a simplified system description. Our presentation here again shares similarities with Schaub et al. (2019b). As a general theme, we will see that whereas the question of timescale separation was mostly concerned with the eigenvalues of the system matrix, symmetries are closely related to a particular structure of the eigenvectors.

5.1 Equivalence Classes and Symmetries of Nodes

To introduce the general idea of how symmetries can simplify the analysis of a dynamical system on a network, let us once more consider the specific examples of a consensus process $\dot{\boldsymbol{x}} = -\boldsymbol{L}\boldsymbol{x}$. Note that in order to write down this dynamical system, we have associated to each node a particular label $i \in \{1, \ldots, n\}$, that is, we have created a labeled graph whose algebraic representation is provided by the Laplacian $\boldsymbol{L}$ (see Figure 7).

Let us now assume that there exists a relabelling of the nodes, defined as the permutation function $\gamma\colon \{1, \ldots, n\} \to \{1, \ldots, n\}$, which leaves $\boldsymbol{L}$ invariant. In the simplest case where we merely switch the labels of two nodes i and j, this invariance means that the connections of i and j to any other node are the same (i.e., nodes i and j have the same structural connectivity patterns). Instead of simply exchanging two nodes i and j we may consider more general permutations γ, to which we associate a permutation matrix $\boldsymbol{\Gamma}$ with entries $\Gamma_{\gamma(i),i} = 1$ and zero otherwise. Exploiting the fact that permutation matrices are orthogonal, that is $\boldsymbol{\Gamma}\boldsymbol{\Gamma}^\top = \boldsymbol{I}$, we can express a general symmetry-induced invariance algebraically as

$$\boldsymbol{L} = \boldsymbol{\Gamma}\boldsymbol{L}\boldsymbol{\Gamma}^\top \quad \Leftrightarrow \quad \boldsymbol{L}\boldsymbol{\Gamma} = \boldsymbol{\Gamma}\boldsymbol{L}. \tag{5.1}$$

Now assume that at any point in time t_0, we observe that the state variables associated with two interchanged nodes have identical values, such that $x_i(t_0) = x_{\gamma(i)}(t_0)$. For instance, this scenario may be of interest in the context of consensus dynamics, opinion formation, or synchronisation processes on networks (Bullo, 2019; Proskurnikov & Tempo, 2017). From our argument above, we can conclude that the state variables will evolve identically for all times thereafter, that is, we have $x_i(t) = x_{\gamma(i)}(t)$ for $t > t_0$. This can be proven by noting that at time t_0 we will have $\boldsymbol{x} = \boldsymbol{\Gamma}\boldsymbol{x}$, and thereafter the evolution is governed by the differential equation

$$\dot{\boldsymbol{x}} = -\boldsymbol{L}\boldsymbol{x} = -\boldsymbol{L}\boldsymbol{\Gamma}\boldsymbol{x} = -\boldsymbol{\Gamma}\boldsymbol{L}\boldsymbol{x} = \boldsymbol{\Gamma}\dot{\boldsymbol{x}}. \tag{5.2}$$

Hence, if we can identify a permutation of the node labels that leaves the system matrix $\boldsymbol{L}$ invariant, we can consider the nodes whose labels were interchanged as an equivalence class of nodes that follow the same equations of motion once their state variables agree at a given time. This corresponds to the idea of node roles, as discussed in Section 3.3, and more specifically to so-called automorphic equivalence (Brandes, 2005, Chapter 9). However, here we are concerned with the dynamical consequences of such an equivalence. Specifically, for every set of nodes that can be interchanged via a symmetry, we have just one equation of motion instead of many independent equations, if the node states are equal at some point in time. Thus we can reduce the number of equations and lump together all the nodes that are equivalent to each other. The reduction in complexity of the system description can be quite significant.

We remark that, in contrast to the approximate system description that we obtained when considering timescale separation, the reduced description based on node equivalence classes is exact at all times, provided the nodes with interchanged labels have identical state variables initially. This can also be seen by examining the consequences of Eq. (5.1) for the eigenvectors of $\boldsymbol{L}$. Specifically, let $\boldsymbol{v}$ be an eigenvector of $\boldsymbol{L}$ with a simple eigenvalue λ (i.e., an eigenvalue with algebraic multiplicity 1). Then $\boldsymbol{\Gamma}\boldsymbol{v}$ will be an eigenvector of $\boldsymbol{L}$ with the same eigenvalue, since $\boldsymbol{L}\boldsymbol{\Gamma}v = \boldsymbol{\Gamma}\boldsymbol{L}\boldsymbol{\Gamma}^\top\boldsymbol{\Gamma}\boldsymbol{v} = \boldsymbol{\Gamma}\boldsymbol{L}\boldsymbol{v} = \lambda\boldsymbol{\Gamma}\boldsymbol{v}$. However, as the eigenvector of a simple eigenvalue is unique up to a scaling factor, this implies that eigenvector entries corresponding to the nodes related by symmetry must be the same.

Symmetries and Node Equivalence under Linear Dynamics

As we discussed in Section 3.3, a number of equivalence notions of nodes have been proposed in the literature, especially within the social network analysis literature. In the context of social networks, such equivalences are also often called node roles (Brandes, 2005, Chapter 9), as the position of a node in a social network is typically thought to define its social status, an idea that is also at the core of centrality measures (Wasserman et al., 1994). In light of our earlier discussion, many of these equivalence notions can be seen as considering specific classes of permutations of either the adjacency matrix $\boldsymbol{A}$ or the combinatorial Laplacian $\boldsymbol{L}$. For instance, two nodes i, j are said to be structurally equivalent if there exists a permutation $\gamma_{\text{struct.}}: \mathcal{V} \to \mathcal{V}$ that exchanges *only* the labels i, j, yet leaves the (labeled) adjacency matrix of the graph invariant ($\boldsymbol{A} = \boldsymbol{\Gamma}_{\text{struct.}} \boldsymbol{A} \boldsymbol{\Gamma}_{\text{struct.}}^{\top}$).

However, the above discussion can be generalised beyond the adjacency matrix and the Laplacian matrix to coupling matrices of any general linear dynamics $\dot{\boldsymbol{x}} = \boldsymbol{F}\boldsymbol{x}$. If there is a permutation between the node labels that leaves the system matrix $\boldsymbol{F}$ invariant, these permutations give rise to a set of (dynamic) equivalence classes of the nodes, in which two nodes are considered equivalent if they can be mapped onto each other by such a permutation. We can equivalently think of these equivalence classes as inducing a partition of the nodes into disjoint groups, where two nodes are in the same group if they are in the same equivalence class. Importantly, these equivalence classes are not defined solely by the underlying topology of the network, but by the properties of the linear dynamical process defined on it, as encoded in the system matrix $\boldsymbol{F}$.

5.2 Equitable and Externally Equitable Partitions

Closely related to the symmetry-induced partitions discussed above are so-called equitable partitions, or EPs for short (Godsil & Royle, 2013), and the related concept of externally equitable partitions (EEPs). As we will see, EEPs provide another viewpoint to derive an exact, simplified description of a dynamical (diffusion) process taking place on a network (O'Clery et al., 2013; Schaub et al., 2016). Conceptually, instead of focusing on symmetries of the operator $\boldsymbol{F}$ that drives the dynamical system, we will now focus directly on invariance properties with which we can characterise groups of nodes that are dynamically similar.

In order to define EEPs, we first recall the well-known graph-theoretic notion of an EP (Godsil & Royle, 2013). An equitable partition splits a graph into a partition $\mathcal{P} = \{\mathcal{A}_1, \ldots, \mathcal{A}_C\}$ of non-overlapping groups of nodes, such that for each node i in group $\mathcal{A}_\alpha$, the number of connections to nodes in a group $\mathcal{A}_\beta$ is dependent only on α, β. In other words, nodes inside each group of an equitable partition have the same out-degree with respect to every group. For an EEP, this condition is relaxed, such that all nodes within group $\mathcal{A}_\alpha$ are required merely to have the same number of links to any other group $\mathcal{A}_\beta$ with $\alpha \neq \beta$. We display an example of an EEP of a graph in Figure 8A.

Importantly, for every EEP we can derive an algebraic relation akin to Eq. (5.1). Specifically, consider an EEP of a graph of n nodes into C groups encoded via the $n \times C$ partition indicator matrix $\boldsymbol{H}$. Then the following algebraic relationship holds:

$$\boldsymbol{L}\boldsymbol{H} = \boldsymbol{H}\boldsymbol{L}^\pi, \tag{5.3}$$

where $\boldsymbol{L}^\pi$ is the $C \times C$ Laplacian of the so-called quotient graph induced by $\boldsymbol{H}$:

$$\boldsymbol{L}^\pi = (\boldsymbol{H}^\top \boldsymbol{H})^{-1}\boldsymbol{H}^\top \boldsymbol{L}\boldsymbol{H} = \boldsymbol{H}^+\boldsymbol{L}\boldsymbol{H}. \tag{5.4}$$

Here the $C \times n$ matrix $\boldsymbol{H}^+$ is the (left) Moore–Penrose pseudoinverse of $\boldsymbol{H}$. Note that the above formula simply builds an appropriate quotient Laplacian $\boldsymbol{L}^\pi$ by computing the (normalized) number of edges running between the different groups (cf. Fig. 8). We remark that although the Laplacian $\boldsymbol{L}$ of the original (undirected) graph is symmetric, the Laplacian $\boldsymbol{L}^\pi$ of the quotient graph will generally be non-symmetric. However, it can be shown that the eigenvalues of the quotient graph correspond to a subset of the eigenvalues of the original graph Laplacian and thus the eigenvalues of $\boldsymbol{L}^\pi$ are also real (O'Clery et al., 2013; Schaub et al., 2016).

Note that the quotient graph is merely a coarse-grained version of the original graph: (i) each equivalence class (group) of nodes gets collapsed into one node; (ii) the weights of the links between these new nodes correspond to the number of links each equivalent node has to the other groups in the original graph (Figure 8A). To see this algebraically, observe that multiplying a vector $\boldsymbol{x} \in \mathbb{R}^n$ by $\boldsymbol{H}^\top$ from the left leads to a C-dimensional vector that records the sums over all components of $\boldsymbol{x}$ within each group. Further, $\boldsymbol{H}^\top\boldsymbol{H}$ is a diagonal matrix with the number of nodes per group on the diagonal. Hence, $\boldsymbol{H}^+ = (\boldsymbol{H}^\top\boldsymbol{H})^{-1}\boldsymbol{H}^\top$ can be interpreted as a group averaging operator (O'Clery et al., 2013).

5.2.1 Dynamical Implications of EEPs

The algebraic characterisation of an EEP, Eq. (5.3) implies that the partition indicator matrix $\boldsymbol{H}$ defines an invariant subspace with respect to the Laplacian $\boldsymbol{L}$. Specifically, if we multiply the Laplacian matrix with the indicator matrix $\boldsymbol{H}$, we obtain a linearly rescaled (by $\boldsymbol{L}^\pi$) version of $\boldsymbol{H}$. As all invariant subspaces of $\boldsymbol{L}$ are expressible in terms of the eigenvectors of $\boldsymbol{L}$, it follows that

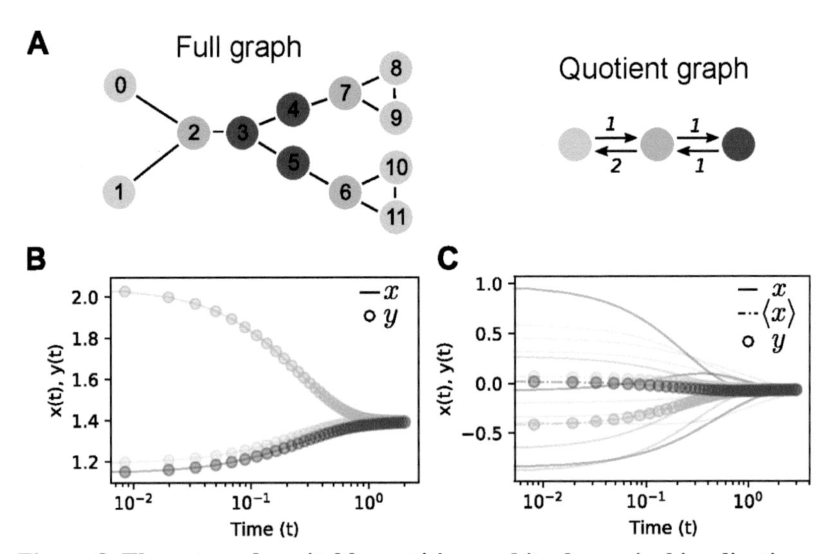

Figure 8 The external equitable partition and its dynamical implications. **A** Left: A graph with $n = 12$ nodes with an external equitable partition into three groups (color coded). Right: The quotient graph associated to the EEP. **B** Invariance of the EEP: The consensus dynamics on the full graph, determined by Eq. (2.15) (lines), from an initial condition $\boldsymbol{x} = \boldsymbol{H}\boldsymbol{y}$ is shown in comparison to the associated quotient dynamics (Eq. (5.5)) governing $\boldsymbol{y}$ (circles). If the states within each group are equal, the dynamics will be described by the dynamics of the quotient graph for all times. **C** Group-averaging dynamics of the EEP: For consensus dynamics, the quotient graph dynamics (circles) also describes the group-averaged dynamics (dash-dotted lines) of the unsynchronised full graph dynamics (lines), as given by Eq. (5.6).

there exists a set of eigenvectors of $\boldsymbol{L}$, such that for each group of the partition, the eigenvector takes the same value for each node. More precisely, there exists a set of C eigenvectors that can each be expressed as $\boldsymbol{v} = \boldsymbol{H}\boldsymbol{v}^{\pi}$ for some vector $\boldsymbol{v}^{\pi} \in \mathbb{R}^{C}$. As it turns out, these vectors $\boldsymbol{v}^{\pi}$ correspond to the eigenvectors of $\boldsymbol{L}^{\pi}$. A further consequence of this fact is that the eigenvalues associated with these eigenvectors (that span the invariant subspace associated to $\boldsymbol{H}$) are shared with $\boldsymbol{L}^{\pi}$. We refer to O'Clery et al. (2013); Schaub et al. (2016) for further discussions on these aspects.

The algebraic properties of an EEP have implications for any Laplacian dynamics $\boldsymbol{L}$. For instance, for consensus dynamics, we obtain the following implications (O'Clery et al., 2013). First, an EEP is consistent with *partial consensus* such that the agreement within any present group is preserved. Specifically, consider an initial state vector $\boldsymbol{x} = \boldsymbol{H}\boldsymbol{y}$ for some arbitrary $\boldsymbol{y}$, such that every node within a group $\mathcal{A}_{\alpha}$ has the same initial value y_{α}. It then follows from the properties of an EEP that the nodes inside each group remain identical *for all times*, and that dynamics of the group variables $\boldsymbol{y}$ can be exactly described by the quotient graph:

$$\boldsymbol{x}(t) = \boldsymbol{H}\boldsymbol{y}(t) \quad \text{with} \quad \dot{\boldsymbol{y}} = -\boldsymbol{L}^{\pi}\boldsymbol{y}, \tag{5.5}$$

which can be directly derived from Eq. (5.3). The dynamical invariance induced by the EEP thus provides a simpler model of the system in the same vein as the symmetry reduction discussed before.

A second consequence of the presence of an EEP is that the dynamics of the group-averaged states $\langle \boldsymbol{x} \rangle$ is exactly described by the quotient graph:

$$\frac{d\langle \boldsymbol{x} \rangle}{dt} = -\boldsymbol{L}^{\pi}\langle \boldsymbol{x} \rangle \quad \text{where} \quad \langle \boldsymbol{x} \rangle := \boldsymbol{H}^{+}\boldsymbol{x}. \tag{5.6}$$

This can be shown by noting that there exists a similar relationship to Eq. (5.3) between the group averaging operator $\boldsymbol{H}^{+}$ and the Laplacians of the original and quotient graphs:

$$\boldsymbol{H}^{+}\boldsymbol{L} = \boldsymbol{L}^{\pi}\boldsymbol{H}^{+}. \tag{5.7}$$

As a consequence, the group-averaged dynamics is also governed by the lower dimensional quotient Laplacian (Figure 8C). Thus, if we are interested only in the averages over the groups of an EEP, we can reduce our model significantly.

Finally, a third implication of the EEP structure relates to the dynamical system with inputs. It can be shown (O'Clery et al., 2013) that all the results for the autonomous consensus dynamics with no inputs can be equivalently rephrased for the system with inputs:

$$\dot{\boldsymbol{x}} = -\boldsymbol{L}\boldsymbol{x} + \boldsymbol{u}(t), \tag{5.8}$$

when the input $\boldsymbol{u}(t) = \boldsymbol{H}\boldsymbol{v}(t)$, $\boldsymbol{v}(t) \in \mathbb{R}^{C}$ is consistent with the cells of an EEP. In that case, the nodes inside each cell remain identical for all times, as in Eq. (5.5).

Remark: While we have focused here on the implications of an EEP for linear consensus dynamics, invariant partitions like the EEP play a similar role for other linear and *nonlinear* dynamics (e.g., Kuramoto synchronisation). See Schaub et al. (2016) for an extended discussion including synchronisation dynamics, as well as dynamics on signed networks.

5.3 Stochastic Symmetries and Equivalences

In the previous sections, we have discussed symmetries and (externally) equitable partitions of a graph, and their implications for linear dynamics.

Rather than applying these concepts to a specific graph, we may also employ them in the context of *random* graphs, as we will discuss in this section.

As a concrete example, consider again the stochastic blockmodel as introduced in Section 3.3. Neglecting the issue of self-loops, based on the definition of the stochastic block model (cf. Eq. (3.9)) it can be shown that the *expected* adjacency matrix of the SBM can be written as

$$\mathbb{E}[\boldsymbol{A}] = \boldsymbol{H}\boldsymbol{\Omega}\boldsymbol{H}^\top, \tag{5.9}$$

where $\boldsymbol{H}$ is the partition indicator matrix, as described previously and $\boldsymbol{\Omega}$ is the affinity matrix of the SBM. Equation (5.9) is simply a consequence of the stochastic equivalence of the nodes: indeed, for any permutation $\boldsymbol{\Gamma}$ that maps nodes within the blocks onto each other, we will have $\boldsymbol{\Gamma}\mathbb{E}[\boldsymbol{A}]\boldsymbol{\Gamma}^\top = \mathbb{E}[\boldsymbol{A}]$. Following our discussion at the end of Section 5.1, this implies that the expected adjacency matrix has piecewise-constant eigenvectors, since all nodes within a block are related by symmetry.

It turns out that this knowledge about the eigenvectors of the *expected* adjacency matrix can also help us to understand the eigenvectors of a network sampled from an SBM. For this we will turn to the Davis–Kahan theorem, which enables us to relate the eigenvectors of a matrix $\boldsymbol{A}$ to the eigenvectors of a perturbed version $\hat{\boldsymbol{A}}$ of that matrix. We state this theorem here in the form given in Yu, Wang, and Samworth (2015):

> **Theorem 2 (Davis–Kahan)** *Let $\boldsymbol{M}, \hat{\boldsymbol{M}} \in \mathbb{R}^{n\times n}$ be symmetric, with eigenvalues $\lambda_1 \geq \ldots, \lambda_n$ and $\hat{\lambda}_1, \ldots, \hat{\lambda}_n$, respectively. Fix r, s such that $1 \leq r \leq s \leq n$ and assume that $\Delta_\lambda := \min(\lambda_{r-1} - \lambda_r, \lambda_s - \lambda_{s-1}) > 0$, where we set $\lambda_0 = \infty$ and $\lambda_{n+1} = -\infty$. Let $d = s - (r-1)$ and let $\boldsymbol{V} = [\boldsymbol{v}_r, \ldots, \boldsymbol{v}_s] \in \mathbb{R}^{n\times d}$ and $\hat{\boldsymbol{V}} = [\hat{\boldsymbol{v}}_r, \ldots, \hat{\boldsymbol{v}}_s] \in \mathbb{R}^{n\times d}$ be the matrices of (orthonormal) eigenvectors corresponding to the eigenvalues $\lambda_r, \ldots, \lambda_s$ of $\boldsymbol{M}$ and the eigenvalues $\hat{\lambda}_r, \ldots, \hat{\lambda}_s$ of $\hat{\boldsymbol{M}}$, respectively. Then, there exists an orthogonal matrix $\boldsymbol{O}$ such that:*

$$\|\hat{\boldsymbol{V}} - \boldsymbol{V}\boldsymbol{O}\|_F \leq \frac{\sqrt{8d}\,\|\boldsymbol{M} - \hat{\boldsymbol{M}}\|_2}{\Delta_\lambda} \tag{5.10}$$

Let us translate these results into our context here. We aim to apply the Davis–Kahan theorem with the matrices $\boldsymbol{M} = \mathbb{E}[\boldsymbol{A}]$ and $\hat{\boldsymbol{M}} = \boldsymbol{A}$ corresponding to the expectation of the adjacency matrix under the SBM and one sample from the model. The left-hand side of Eq. (5.10) would then correspond to a distance between the eigenvectors of $\mathbb{E}[\boldsymbol{A}]$ and $\boldsymbol{A}$, modulo a rotation/reflection of those eigenvectors via the matrix $\boldsymbol{O}$.[13] To obtain a bound on this distance we need to ensure that the right-hand side is small, which can be guaranteed if two requirements are fulfilled. First, the SBM needs to have parameters such that the nonzero eigenvalues of the expected adjacency matrix $\mathbb{E}[\boldsymbol{A}]$ are well-enough separated from zero (i.e., Δ_λ will be large enough). Second, we require that the difference $\|\boldsymbol{A} - \mathbb{E}[\boldsymbol{A}]\|$ between any sampled adjacency matrix and its expectation is comparably small with high probability. Specifically, this term needs to be much smaller compared to Δ_λ (note that both terms may in general depend on n). Using concentration of measure techniques, as mentioned in Section 4.3.2, such a matrix concentration result for $\|\boldsymbol{A} - \mathbb{E}[\boldsymbol{A}]\|$ can indeed be established for the SBM (Le, Levina, & Vershynin, 2017; Lei et al., 2015), provided the expected degrees of the nodes are not too small.

Given these two conditions are fulfilled, we can then invoke the Davis–Kahan theorem (Stewart, 2001; Yu et al., 2015) to conclude that the eigenvectors of a sampled adjacency matrix $\boldsymbol{A}$ from the stochastic blockmodel will remain close to the eigenvectors of the expected adjacency matrix with high probability. Since the expected adjacency matrix has piecewise-constant eigenvectors, the eigenvectors of the sampled adjacency matrix will thus be approximately piecewise constant. In particular, this also demonstrates that the dominant eigenvectors of the adjacency matrix can be used for spectral clustering, which provides support for the use of spectral methods for community detection or graph partitioning, as introduced in Section 3.2.

The connection between equitable partitions and SBMs provides a simple example for the utility of considering symmetry properties (or node-equivalences, respectively) of the expected adjacency matrix of (edge-independent) random graph models. An alternative way to reach this conclusion is to observe that the expected adjacency matrix under the SBM can be split according to an equitable partition. However, the requirements of an SBM are in fact stronger than imposing that there exists an EP for the expected adjacency matrix of the model. We may thus introduce the concept of a stochastic equitable partition sEP (Schaub & Peel, 2020), which says that the expected adjacency matrix can be equitable partitioned (see Figure 9 for an illustration). Importantly, the number of expected adjacency matrices with an sEP with C groups is generally larger than the number of SBMs with the same number of groups.

[13] Note that the term $\boldsymbol{O}$ is unavoidable here, in general, as only the eigenspaces are uniquely defined, but not the specific eigenvectors. Indeed even a single normalized eigenvector is unique only up to sign.

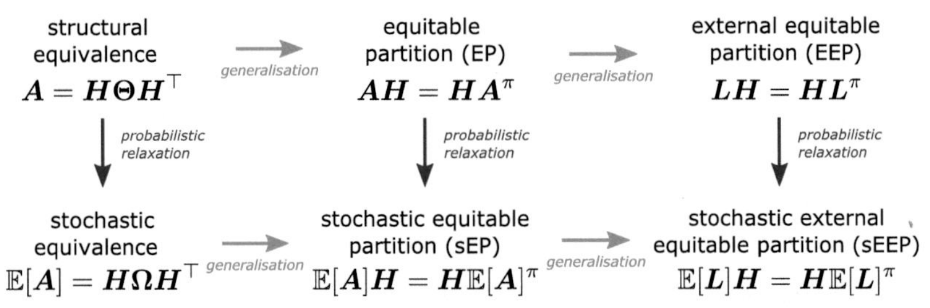

Figure 9 Overview of network partition equivalence relationships. The top line describes partitions (represented by a partition indicator matrix $\boldsymbol{H}$) into groups of equivalent nodes in a given graph (represented by an adjacency matrix $\boldsymbol{A}$ or a Laplacian $\boldsymbol{L}$). The bottom line presents the corresponding probabilistic relaxation in which the equivalence relation is considered in terms of the expected adjacency matrix $\mathbb{E}[\boldsymbol{A}]$ over the ensemble of networks generated by a random graph model. (Note that for simplicity we allow for graphs with self-loops in the algebraic expressions of structural and stochastic equivalence.) *Structural equivalence:* Nodes are equivalent if they link to the same neighbors. Here $\boldsymbol{\Theta}$ is a $\{0, 1\}$ matrix. *Stochastic equivalence:* Nodes are structurally equivalent in expectation. *Equitable partition:* Nodes are equivalent if they have the same number of links to equivalent nodes. *Stochastic equitable partition:* The partition is an EP *in expectation.* *Externally equitable partition:* Nodes are equivalent if they have the same number of links to equivalent nodes, outside their own group. *Stochastic externally equitable partition:* The partition is an EEP *in expectation*. Figure adapted from Schaub & Peel (2020).

Rather than considering only sEPs, we may further consider stochastic EEPs, which we can define analogously to sEPs by requiring that the expected adjacency matrix of a random graph model can be partitioned according to an EEP. This construction has similar implications for the eigenvectors of the graph Laplacian of a graph sampled from this model, as discussed for sEPs in terms of the adjacency matrix. This line of thought opens many other possibilities and generalisations of the concept of stochastic equivalences. For instance, it enables us to provide an interesting characterisation of the concept of hierarchical random graph. As discussed in Schaub and Peel (2020), we may conceptualise a hierarchical modular structure as one that exhibits a sEP for each hierarchical partition. Moreover, since EEPs are also characterised by piecewise constant eigenvectors, we can make similar arguments as in the case of the SBM and can derive that the eigenvectors of sampled adjacency matrices will be approximately constant on each node-group under suitable assumptions.

5.4 Differences and Relationships between EEPs and Timescale Separation

Let us discuss briefly the difference between the presence of an EEP and timescale separation in a network. Both concepts can be related to strictly invariant subspaces (EEPs) or almost invariant subspaces (timescale separation) in the dynamics. However, the link between structure and dynamics that each of them represents is different. In fact, the notions of EEP and timescale separation are distinct but not mutually exclusive.

The presence of an EEP is related to symmetries in the graph, which translate into the fact that a set of Laplacian eigenvectors have components that are constant on each cell in the graph. These eigenvectors can be associated to *any eigenvalue* of the graph (i.e., these eigenvectors can be fast or slow eigenmodes). In broad terms, for an EEP the piecewise constant structure of the eigenvectors with respect to the groups is important, but the eigenvalues themselves are not relevant. This notion is therefore different from the timescale separation discussed in Section 4.1, where the defining criterion focuses on the eigenvalues – more precisely, on the existence of gaps between eigenvalues that separate them into groups associated with different timescales.

In our particular example in Figure 5, the associated eigenvectors were indeed approximately piecewise constant on each group (i.e., on each block of nodes). Hence, in this case both the approximate EEP structure and the timescale separation are well aligned. In fact, in the case of an approximate EEP structure, we actually need some kind of eigenvalue separation to invoke the Davis–Kahan theorem. However, timescale separation and EEP structure may not always be aligned, and we can have EEPs associated to fast eigenmodes as well. Likewise, the eigenvectors corresponding to the slowest timescales in a system with timescale separation do not have to be exactly piecewise constant and correspond to an EEP in general (Schaub et al., 2015; Schaub et al., 2012).

5.5 Further Discussion and References

The importance of symmetries for dynamics is a classical topic and has been considered in the context of network dynamics, for example, by Golubitsky and Stewart (2006, 2015); Stewart, Golubitsky, and Pivato (2003). For more thorough discussions on the relations between EPs, EEPs, automorphic equivalence, and graph isomorphism, we refer to Chan and Godsil (1997); Grohe et al. (2014); Grohe and Schweitzer (2020). Important references on how to use these concepts in the context of the analysis of networks and network dynamics include Cardoso, Delorme, and Rama (2007); Egerstedt et al. (2012); Pecora et al. (2014); Sanchez-Garcia (2018).

We have emphasised here the conceptual differences between timescale separation and presence of EEPs in a network. An interesting venue of research would be to explore further the intersection between these concepts. For instance, it would be of interest to develop metrics that can capture both how close a given partition is to being a EEP and how dominant its associated eigenvectors are. First steps in this direction include information theoretic approaches focusing on Markov dynamics on networks (Faccin, Schaub, & Delvenne, 2018, 2020).

A different direction for future research would be to generalise the concept of an EEP. As discussed, we can associate an EEP with an invariance with respect to the graph Laplacian (see Eq. (5.3)). Using this algebraic characterisation as a starting point, we may also define partitions that are invariant (and in this sense 'equitable') with respect to other matrices. For instance, we may define a signed external equitable partition with respect to the so-called signed Laplacian (Schaub et al., 2016), but many other choices are possible as well and would warrant further exploration.

6 Dynamical Methods for Assortative Communities

In the previous two sections, we have discussed how modular network structure can affect dynamics and, in particular, diffusion dynamics on networks. Specifically, we have seen in Section 4 that strong assortative communities can induce a timescale separation, or equivalently, a separation between the eigenvalues of the system matrix governing the dynamics. In the following sections, we will consider the reverse direction and ask how we can utilise a dynamical process to detect communities within a network, focusing on assortative community structure. The intuitive idea is that as assortative communities can induce a timescale separation, we can search for particular subparts in a network in which a diffusion process will be trapped for an (unexpectedly) long time. This idea underpins a number of successful community detection methods, often called dynamical or flow-based methods, including the WalkTrap algorithm of Pons and Latapy (2005) or the so-called map-equation framework by Rosvall and Bergstrom (2008). Here we will concentrate on the Markov stability method (Delvenne et al., 2013; Delvenne, Yaliraki, & Barahona, 2010, Lambiotte, Delvenne, & Barahona, 2014; Schaub et al., 2012), as it is conceptually closest to our previous discussions and moreover provides a framework under which we can understand a number of other well-known community detection algorithms from a dynamical lens. We provide a brief discussion of alternative approaches and related ideas at the end of the section.

6.1 Basics of Markov Stability

In this section, we present the Markov stability framework (Delvenne et al., 2010), which enables us to define quality functions for community detection based on random-walk dynamics. Strikingly, it can be shown that Markov stability considers the number of methods used in network analysis (Delvenne et al., 2013) which have been developed without any dynamical process in mind. In particular, Markov stability provides an alternative interpretation of the Newman–Girvan modularity (3.3), which can explain some of the limitations of modularity from a dynamical perspective and remedy them through the introduction of a resolution parameter associated to the dynamics.

Consider an ergodic random-walk process, in discrete or continuous time, on a network. The *Markov stability* of a partition of the graph at time t is defined as the difference of two probabilities (Delvenne et al., 2010; Lambiotte et al., 2014): first, the probability of a stationary random walker to be in the same community at time 0 and at time t; second, the probability of a stationary random walker to be in the same community at time 0 and in the limit $t \to \infty$. Thus, Markov stability can be interpreted as measuring the persistence of a random-walk process inside the communities of the partition. Within a given timescale t, Markov stability is large when it is unlikely for the random walker to have left the community in which it started. As we will see, Markov stability can also be understood as the autocovariance of a signal encoding the sequence of communities visited by the random-walk process. Note that it can be defined for different random-walk processes, each one giving rise to a different quality function and, in principle, to a different optimal partition of the same network.

Let us remark that for an ergodic random walk (i.e., a random walk on an aperiodic, strongly connected graph) the second, asymptotic probability appearing in Markov stability is equivalent to the probability of two independent walkers to be in the same community at time t by chance, as the information of the initial conditions is lost for long times. In the following, we will exclusively consider this case of aperiodic, strongly connected graphs. Importantly, if the underlying graph does not satisfy those properties, we can always construct such a dynamics by incorporating a 'teleportation' probability at each step of the random walk (Brin & Page, 1998; Lambiotte & Rosvall, 2012), that is, a small probability to randomly jump to any other node in the graph.

We have discussed Markov stability on an abstract and general level so far. To be more concrete, let us first concentrate on a discrete-time random walk on an undirected network. The time evolution of the probability $p_i(t)$ to be located on node i at time t is governed by the equation

$$\boldsymbol{p}^\top(t+1) = \boldsymbol{p}^\top(t)\boldsymbol{T}, \tag{6.1}$$

where $\boldsymbol{T} = \boldsymbol{K}^{-1}\boldsymbol{A}$ is the transition matrix of the associated Markov chain (2.22). Under the assumptions of a connected and non-bipartite graph, the process converges to a unique stationary distribution,

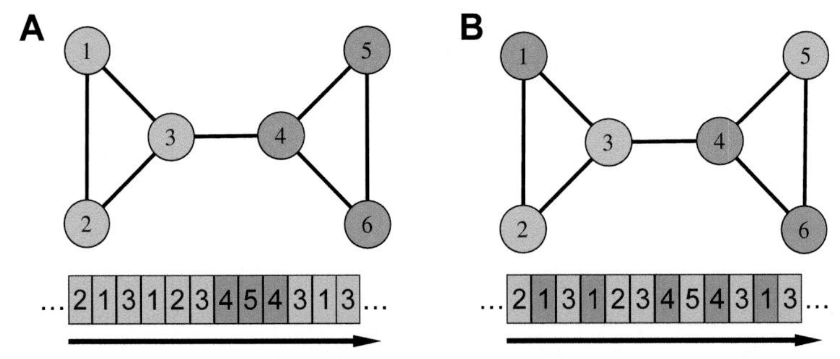

Figure 10 Markov stability and random walks. Given a partition of a graph, here illustrated by two colours, Markov stability is defined by the sequence of communities visited by the random-walk process. Intuitively, for a good partition as in A, the random walker will persist for long times inside a community before escaping it. Markov stability captures the persistence of a random walker at a timescale t via its clustered covariance matrix.

$$\boldsymbol{\pi}^\top = \boldsymbol{k}^\top/2m, \tag{6.2}$$

where $\boldsymbol{k}$ is the vector of degrees in the network.

Based on this process, we define the *clustered autocovariance* matrix of the diffusion process at time t as follows (see Figure 10 for an illustration). Given a partition of a network, encoded as before by the indicator matrix $\boldsymbol{H}$, we assign a different real value X_α ($\alpha = 1, \ldots, C$) to the vertices of each of the C communities, such that each node within the same community is assigned the same value X_α. We now consider the sequence of values $X(t)$ elicited by a random-walk process on the network, assuming that the random walk has been initialised in its stationary state at time $t = 0$. The autocovariance of this process evaluated over a period of time t is:

$$\operatorname{cov}[X(0)X(t)] = \mathbb{E}[X(0)X(t)] - \mathbb{E}[X(0)]\mathbb{E}[X(t)], \tag{6.3}$$

where $\mathbb{E}[X(t)]$ is the expectation of the random variable $X(t)$. For a discrete-time random walk, this autocovariance is given by

$$\operatorname{cov}[X(0)X(t)] = \boldsymbol{X}^\top \boldsymbol{R}(t, \boldsymbol{H})\boldsymbol{X}, \tag{6.4}$$

where $\boldsymbol{X}$ is the $1 \times C$ column vector of labels assigned to the C communities and where

$$\boldsymbol{R}(t, \boldsymbol{H}) = \boldsymbol{H}^\top \left[\boldsymbol{\Pi} \boldsymbol{T}^t - \boldsymbol{\pi}\boldsymbol{\pi}^\top\right] \boldsymbol{H} \tag{6.5}$$

is by definition the $C \times C$ clustered covariance matrix. In this last expression, $\boldsymbol{\Pi} = \operatorname{diag}(\boldsymbol{\pi})$ is a diagonal matrix encoding the stationary distribution of the random walk ($\boldsymbol{\pi}^\top = \boldsymbol{\pi}^\top \boldsymbol{T}$).

Observe that the clustered autocovariance matrix $\boldsymbol{R}(t, \boldsymbol{H})$ does not depend on the arbitrary values X_α used to encode the communities. By construction, $(\boldsymbol{\Pi T}^t)_{ij}$ measures the flow of probability from node i to node j in t steps, starting from the stationary distribution of the random walk. Due to the multiplication by the indicator matrices, the term $[\boldsymbol{H}^\top \boldsymbol{\Pi T}^t \boldsymbol{H}]_{\alpha\beta}$ thus measures the flow of probability between any two communities $\mathcal{A}_\alpha$ and $\mathcal{A}_\beta$ over time t. Moreover, as we have assumed that the dynamics is ergodic, the probability to arrive on node j becomes independent of its initial state in the long time limit:

$$\lim_{t\to\infty} (\boldsymbol{\Pi T}^t) = \boldsymbol{\pi}\boldsymbol{\pi}^\top. \tag{6.6}$$

Hence, the second term in the clustered covariance, $\boldsymbol{H}^\top \boldsymbol{\pi}\boldsymbol{\pi}^\top \boldsymbol{H}$, describes the flow of probability between two communities as $t \to \infty$. Note that this also implies that all the elements of $\boldsymbol{R}(t, \boldsymbol{H})$ will converge to zero as $t \to \infty$, irrespectively of the partition considered.

In general, the (α, β) entry of the $C \times C$ matrix $\boldsymbol{R}(t, \boldsymbol{H})$ describes the probability that a random walker will be at community $\mathcal{A}_\alpha$ at time zero and community $\mathcal{A}_\beta$ at time t, minus the probability of these events happening by chance at the stationary state. Intuitively, there is a strong assortative community structure over a timescale t, if the probability flows are contained within the communities, hence concentrating high values on the diagonal of $\boldsymbol{R}(t, \boldsymbol{H})$. Accordingly, the Markov stability for discrete-time random walks is defined via the trace of the clustered autocovariance matrix (Delvenne et al., 2013, 2010):

$$r(t, \boldsymbol{H}) = \min_{0\leq t\leq s} \operatorname{Tr}[\boldsymbol{R}(s, \boldsymbol{H})] \approx \operatorname{Tr}[\boldsymbol{R}(t, \boldsymbol{H})], \tag{6.7}$$

in line with the abstract definition given at the beginning of this section.[14] The Markov stability $r(t, \boldsymbol{H})$ of a partition defined via Eq. (6.7) can be used to rank partitions of a given graph at different timescales. For every value of t, community detection can thus be performed by optimising $r(t, \boldsymbol{H})$ over the space of all possible partitions, resulting in a sequence of optimal partitions over different time intervals.

The Markov stability $r(t, \boldsymbol{H})$ of a partition (6.7) has connections with several concepts related to community detection and graph theory. For instance, when considered at time $t = 1$, it is straightforward to show that $r(1, \boldsymbol{H})$ is equal to

[14] Note that the minimisation in the above formula ensures generality of the definition. Indeed, in almost bipartite (disassortative) graph, $\operatorname{Tr}\boldsymbol{R}(t, \boldsymbol{H})$ can oscillate for discrete time-dynamics. We therefore take the lowest point over the interval as the quality function $r(t, \boldsymbol{H})$. The minimisation is, however, not necessary in most cases and, in particular, it can be proven that it is not required in the continuous time case, for which $\operatorname{Tr}\boldsymbol{R}(t, \boldsymbol{H})$ is always monotonically decreasing (see Delvenne et al. (2013) for more discussion).

the Newman–Girvan modularity Q; see Eq. (3.3). This provides an alternative interpretation of modularity in terms of flows of probability instead of density of links inside communities. When $t = 0$, the Markov stability of a partition simplifies into

$$r(0, \boldsymbol{H}) = 1 - \sum_{\alpha=1}^{c} \pi_\alpha^2 =: GS(\boldsymbol{\pi}, \boldsymbol{H}), \tag{6.8}$$

where $\pi_\alpha = \sum_{i \in \mathcal{A}_\alpha} \pi_i$ is simply the probability that a stationary random walker is in community $\mathcal{A}_\alpha$. We may interpret this probability as the volume of the community. Further, $r(0, \boldsymbol{H})$ is equivalent to the Gini–Simpson diversity index (Simpson, 1949) of the probability distribution induced by the partition, but also to the Rényi entropy of order 2 in Information Theory (Rényi et al., 1961) and to the Deridda and Flyvbjerg number in Physics (). Indeed, this quantity has the intuitive properties of an entropy measure, and it is large when the partition is made of many communities of equal volume and is low when it has few and uneven communities.

6.2 Time as (Nonlinear) Resolution Parameter

To further explore the dependency of Markov stability on time, it is insightful to use a continuous-time formulation, and to consider the continuous-time random walk, see Eq. (2.21), whose master equation reads

$$\frac{d}{dt}\boldsymbol{p}^\top = -\boldsymbol{p}^\top \boldsymbol{L}_{\text{rw}}. \tag{6.9}$$

Under the condition that the network is undirected and connected, the process converges to the same unique stationary distribution as the standard discrete time random walk (6.2). Using the formal solution to the continuous time process

$$\boldsymbol{p}^\top = \boldsymbol{p}_0^\top \exp(-t\boldsymbol{L}_{\text{rw}}), \tag{6.10}$$

and an analogous derivation as in discrete time, the Markov stability for a partition of a continuous-time random walk takes the form

$$r(t, \boldsymbol{H}) = \text{Tr}\left[\boldsymbol{H}^\top \left[\boldsymbol{\Pi} \exp(-t\boldsymbol{L}_{\text{rw}}) - \boldsymbol{\pi}\boldsymbol{\pi}^\top\right] \boldsymbol{H}\right]. \tag{6.11}$$

As before, this expression provides a quality function that is parametrically dependent on time.

Observe that the matrix exponential comprises matrix powers of all positive integer exponents, corresponding to walks of all lengths in the graph. These matrix powers are scaled by time, such that larger values of t give more weight to longer walks, corresponding to an exploration of the network at a larger scale. Time thus acts as a *nonlinear* resolution parameter which enables us to change the scale of the preferred communities within the Markov stability framework: short times lead generally to smaller communities, longer time to larger communities.

To understand how time acts as a resolution parameter, it is instructive to consider the behaviour of (6.11) in the limit of small and large times. When $t = 0$, one recovers again the Gini–Simpson index, whose maximisation leads to a partition of n communities, each made of one single node. This solution is expected from our interpretation of the Gini–Simpson index as an entropy, and it provides the finest-grained partition of a network. In order to explore further the limit of small times, we perform a Taylor expansion of (6.11) around $t = 0$ and obtain a linearised version of Markov stability

$$r(t, \boldsymbol{H}) \approx r(0, \boldsymbol{H}) + t \left. \frac{dr(t, \boldsymbol{H})}{dt} \right|_{t=0} = r(0, \boldsymbol{H}) - t\mathrm{Tr}\left(\boldsymbol{H}^\top \frac{\boldsymbol{L}}{2m} \boldsymbol{H}\right) \tag{6.12}$$

or equivalently

$$r(t, \boldsymbol{H}) \approx GS(\boldsymbol{\pi}, \boldsymbol{H}) - t\mathrm{Cut}, \tag{6.13}$$

where $\mathrm{Cut} = \mathrm{Tr}[\boldsymbol{H}^\top \boldsymbol{L} \boldsymbol{H}]/2m$ simply counts the fraction of the edges between the communities. This expression provides an interpretation of Markov stability as the competition between two objectives. The second term is minimised if we have a small number of links between communities, which favours aggregating the nodes into large groups of vertices. The first term, the Gini–Simpson index, favours, however, a large number of equally-sized, balanced communities. The relative weight between both objectives is modulated as the Markov time t increases. Starting from $t = 0$, where the assignment of every node to its own community is optimal, the optimisation of $r(t, \boldsymbol{H})$ leads to larger and larger communities as time increases. Interestingly it can also be shown that Eq. (6.13) is equivalent, up to a multiplicative constant, to the Potts model heuristic proposed in Reichardt and Bornholdt (2006), which provides a parametric generalisation of modularity. Note that while this correspondence holds for the *linearisation* of the Markov stability of a partition, in general the time parameter will act non-linearly; see Delvenne et al. (2013); Schaub et al. (2019a) for further discussion.

In the limit $t \to \infty$, making use of the spectral decomposition of $\boldsymbol{L}_{\mathrm{rw}}$, stability simplifies to

$$r(t, \boldsymbol{H}) \approx \frac{1}{2m} e^{-\lambda_2 t} \sum_{\alpha=1}^{c} \sum_{i,j \in \mathcal{A}_\alpha} [\boldsymbol{u_2}]_i [\boldsymbol{u_2}]_j, \tag{6.14}$$

where it is assumed that the second dominant eigenvalue λ_2 of $\boldsymbol{L}_{\mathrm{rw}}$ is not degenerate and $\boldsymbol{u}_2$ is its corresponding left eigenvector. The Markov stability of a partition $r(t, \boldsymbol{H})$ is therefore maximised for large times by a split of the network into two communities, defined by the signs of the entries of the eigenvector $\boldsymbol{u}_2$, which is also known as the normalised Fiedler eigenvector. Note that the solution obtained in this way is the same as in the graph partitioning problem of Section 3.2, except that now this split is not the result of an approximation but is exact for sufficiently large values of the time parameter.

The above results highlight that Markov stability provides a rich framework for community detection, in which the timescale of the diffusive process acts as an intuitive resolution parameter, that allows us to uncover the multiscale structure of the network by tuning the characteristic size of the communities in the optimal partition. However, the practical application of the method still involves at least two non-trivial steps.

First, efficient algorithms are needed to optimise Markov stability. Thankfully, it is always possible to rewrite the Markov stability for a given random-walk process as the modularity of another weighted, symmetric network. Any modularity maximisation algorithm can therefore be used for Markov stability optimisation. For the continuous-time, random-walk process considered in this section, for instance, the Markov stability $r(t, \boldsymbol{H})$ of a network with adjacency $\boldsymbol{A}$ is equivalent to the modularity of a network with adjacency matrix $\boldsymbol{Y}(t) = \boldsymbol{\Pi} \exp(-t\boldsymbol{L}_{\mathrm{rw}})$ which encodes the flow of probability in a period of time t between two nodes. Note that the adjacency matrix $\boldsymbol{Y}(t)$ depends explicitly on time and is associated to a weighted network that becomes increasingly dense over time. In addition to the computational cost for the matrix exponential, this significantly increases the storage cost of the network and limits the efficiency of methods designed for modularity optimisation of sparse networks. In situations when these limitations become prohibitive (typically for network sizes on the order of 10^5 nodes), a practical solution is to turn to the linearised version of stability, whose optimisation can be performed with minor modifications of modularity optimisation techniques.

Second, the optimisation of Markov stability across time leads to a sequence of partitions that are optimal at different timescales. This leaves us with the problem of selecting relevant timescales for our description. This is a well-known challenge for multi-resolution methods that may be addressed by considering the robustness of the optima obtained at different values of time (Lambiotte et al., 2014). Notions of robustness are often considered when dealing with NP-hard optimisations and aim to capture the ruggedness of the landscape of the quality function to be optimised (Good et al., 2010). In this

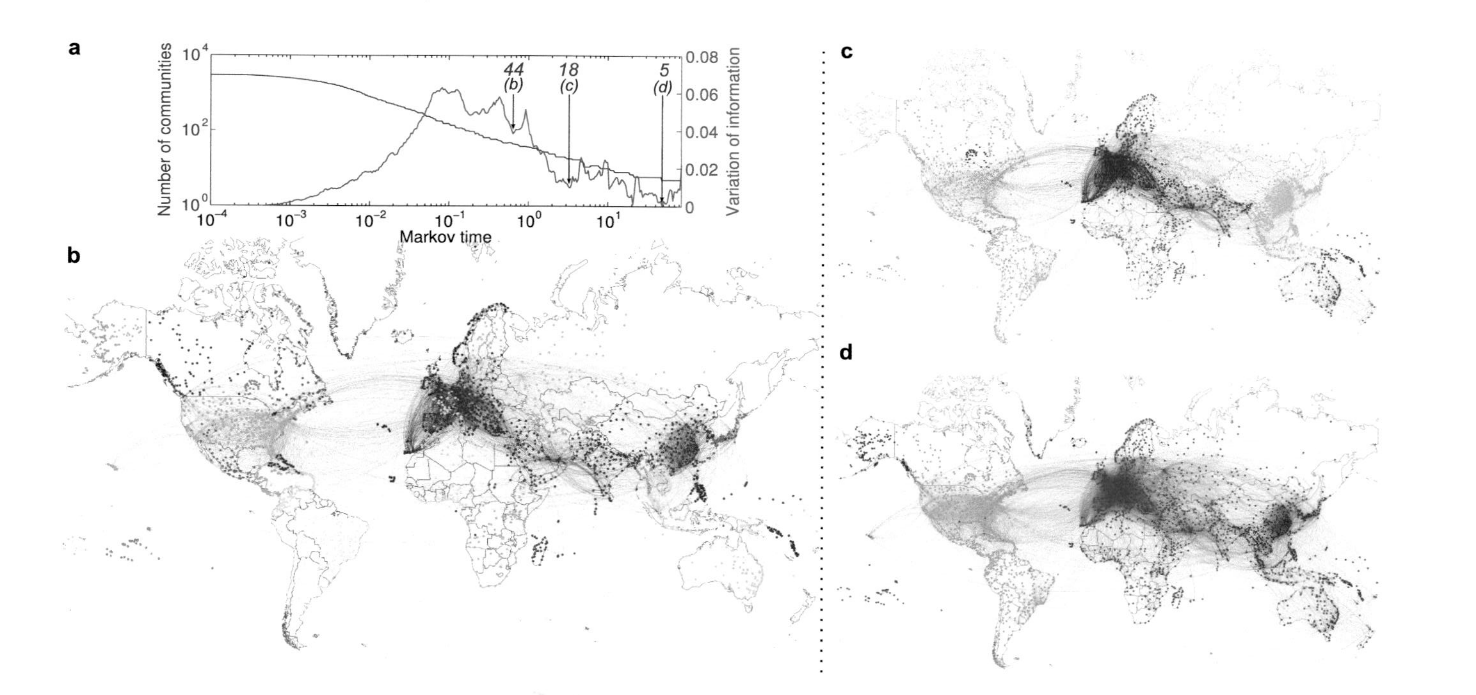

Figure 11 Flow communities at multiple scales in an airport network. The network is made of $n = 2{,}905$ nodes and $m = 30{,}442$ directed edges, whose weight encodes the number of flights between airports. Representative partitions are selected at dips in the normalised variation of information and identify different levels of resolution, here with (b) 44, (c) 18, and (d) 5 communities. Figure reproduced from Lambiotte et al. (2014) with permission.

context, we determine the significance of a given optimum of Markov stability by estimating how it is affected by the addition of noise, for instance by adding small perturbations to the network, or by making small modifications to the optimisation algorithm. A partition is said to be robust if such perturbations have little effect on the outcome and the perturbed result remains close to the unperturbed one. In order to measure the similarity between two partitions $\mathcal{P}_1$ and $\mathcal{P}_2$, a popular choice is the normalised variation of information (Meilă, 2007),

$$\hat{V}(\mathcal{P}_1, \mathcal{P}_2) := \frac{2}{H(\mathcal{P}_1, \mathcal{P}_2) - H(\mathcal{P}_2) - H(\mathcal{P}_1)} \log n, \tag{6.15}$$

where $H(\mathcal{P}_i)$ is the entropy of the partition $\mathcal{P}_i$:

$$H(\mathcal{P}_i) := -\sum_{\alpha=1}^{c} \frac{n_\alpha}{n} \log\left(\frac{n_\alpha}{n}\right), \tag{6.16}$$

where n_α is the number of nodes in group $\mathcal{A}_\alpha$. Analogously we define the joint entropy $H(\mathcal{P}_i, \mathcal{P}_j)$ via:

$$H(\mathcal{P}_i, \mathcal{P}_j) := -\sum_{\alpha=1,\beta=1}^{c_1,c_2} \frac{n_{\alpha\beta}}{n} \log\left(\frac{n_{\alpha\beta}}{n}\right), \tag{6.17}$$

where $n_{\alpha\beta}$ is the number of nodes that is both in group $\mathcal{A}_\alpha$ in partition $\mathcal{P}_1$ and in group $\mathcal{A}_\beta$ in partition $\mathcal{P}_2$. The normalised variation of information $\hat{V}(\mathcal{P}_1, \mathcal{P}_2) \in [0, 1]$ has the desirable property to be a metric in the space of partitions and thus to vanish only when the two partitions are identical. Equipped with this similarity measure, we can characterise the robustness of the optimal partition for each value of time. We then select partitions at timescales which have local minima in the normalised variation of information (i.e., are associated to a higher level of robustness). As an illustration, we show in Figure 11 the result of such an analysis for a weighted network of airport connections. Relevant structure can be found at different resolutions, revealing different levels of geographical and sociopolitical groupings.

6.3 Flow-Based versus Structure-Based Methods in Directed Networks

Many methods for community detection in networks are combinatorial, in the sense that they are based on counting edges inside and between groups of nodes. This is the case for modularity for instance, which counts the number of edges inside communities and compares this expected count in a null model. In contrast, flow-based methods such as Markov stability aim to quantify the effect of the network topology on the flow of a diffusion process on the network. As we have seen above, combinatorial and flow-based methods may be related, and even be equivalent, in the case of undirected networks. However, this equivalence often breaks down when the network is directed.

Markov stability naturally extends to directed networks, for instance by considering the random walk defined by Eq. (6.1) with a transition matrix given by $\boldsymbol{T} = \boldsymbol{K}_{\text{out}}^{-1}\boldsymbol{A}$, where $\boldsymbol{K}_{\text{out}}$ is a diagonal matrix with the out-degrees of each node. Under the assumption that the directed network is strongly connected and aperiodic,[15] the process converges to a unique stationary distribution $\boldsymbol{\pi}$ defined as the dominant left eigenvector of the transition matrix. In contrast to undirected networks, this stationary distribution does not only depend on the degree of each node, but captures the global connectivity patterns in the network and is obtained from a combination of walks of all length. If we include a teleportation probability in the process, this stationary distribution is also equivalent to the PageRank (Gleich, 2015), which is a popular measure of centrality for (directed) networks.

Note that, accordingly, unlike in the case of undirected networks, the Markov stability of a partition at $t = 1$,

$$r(1, \boldsymbol{H}) = \operatorname{Tr}\left[\boldsymbol{H}^\top \left[\boldsymbol{\Pi}\boldsymbol{T} - \boldsymbol{\pi}\boldsymbol{\pi}^\top\right]\boldsymbol{H}\right], \tag{6.18}$$

differs from the standard expression of modularity (3.3), as well as from its most common generalisation to directed networks

$$Q = \frac{1}{M}\operatorname{Tr}\left[\boldsymbol{H}^\top\left[\boldsymbol{A} - \frac{\boldsymbol{k}_{\text{out}}\boldsymbol{k}_{\text{in}}^\top}{M}\right]\boldsymbol{H}\right], \tag{6.19}$$

where the null model is obtained from a directed version of the soft configuration model (Nicosia et al., 2009). While optimisation of $r(1, \boldsymbol{H})$ leads to partitions with persistent flows of probability within modules, modularity favours partitions with high densities of links. The two quantities also differ in their 'null models' terms, as the importance of a node is captured by its PageRank π_i for Markov stability, and by its local connectivity (k_{in}, k_{out}) in the case of modularity. For these reasons, the optimal partitions of Markov stability and modularity usually provide different, complementary community structures for the same directed network. There is no a priori reason that one solution is better than the other one, as they arise from different perspectives and embody different notions of community. Echoing our discussion in Section 3.3, for a better understanding of a specific real-world network, one should

[15] While this assumption is often not met in real data, if the network under consideration is not strongly connected and aperiodic, we can use the teleportation trick discussed before.

thus clearly identify the network aspects that one seeks to understand, when choosing between a combinatorial and a flow-based method.

Before closing this section, let us note that the optimisation of Markov stability for directed networks can also be performed by standard modularity optimisation algorithms after noting that the Markov stability of a given directed network is equal to the modularity of an undirected network whose adjacency matrix is given by

$$\frac{\mathbf{\Pi} \boldsymbol{T} + (\mathbf{\Pi} \boldsymbol{T})^\top}{2}. \tag{6.20}$$

In this weighted network, the importance of an edge is determined by the stationary flow of probability between the nodes, reinforcing our observation that Markov stability takes flows as the measure of importance in a network.

6.4 Different Dynamics Lead to Different Quality Functions

Until now, we have always considered Markov stability for unbiased discrete-time random walks and their continuation in terms of the random-walk Laplacian. However, these linear processes might not be adequate to properly describe the specific dynamics taking place on a graph under scrutiny. Among the systems where unbiased random walks may be unrealistic, one can think of traffic networks, where a bias is necessary to account for local search strategies and navigation rules. One strength of the Markov stability framework is its generality, as it can be defined for random-walk walk process, thus allowing the user to equip the network with an appropriate dynamical model. A natural choice is, for instance,

$$\frac{d}{dt}\boldsymbol{p}^\top = -\boldsymbol{p}^\top \boldsymbol{L}, \tag{6.21}$$

where $\boldsymbol{L}$ is the combinatorial Laplacian and whose stationary state is uniform

$$\boldsymbol{\pi} = \mathbf{1}^\top / n. \tag{6.22}$$

In that case, an important difference is that the diversity index at $t = 0$ now takes the form

$$r(0) = 1 - \sum_{\alpha=1}^{c} \left(\frac{n_\alpha}{n}\right)^2, \tag{6.23}$$

where n_α is the number of nodes in community $\mathcal{A}_\alpha$. The resulting Markov stability thus favours partitions where the communities have the same number of nodes instead of the same number of edges as in (6.8). This example emphasises the importance of choosing appropriate dynamical processes in order to uncover dynamical communities in networked systems. Interesting alternatives include biased random walks, such as the Ruelle–Bowen walk (Delvenne & Libert, 2011) but also higher-order Markov processes (Salnikov, Schaub, & Lambiotte, 2016), where the trajectories of the walkers may be calibrated on empirical data (Rosvall et al., 2014).

RELATIONS TO MODEL ORDER REDUCTION

Dynamical community detection and reduced-order models, as discussed in Section 4.2, both decrease the dimensionality of a linear dynamical system on a network, but in different ways. To clarify this point, let us assume that a dynamical system on a network exhibits C slow eigenvectors. Then, from a model order reduction perspective, the linear system of n equations for the dynamical process can be reduced to a C-dimensional description in the long time limit. The resulting model, which focuses on the evolution of the dynamics in the subspace spanned by the slow eigenvectors, has a bounded error that shrinks over time. This classical result from linear dynamical systems theory is particularly helpful if we want to construct a coarse-grained description of an autonomous dynamical system. However, from a network perspective, this solution is not entirely satisfying, as the new coordinates (the slow eigenmodes) are not necessarily concentrated on groups of nodes (cf. our discussion in Section 2.3.2), and yet the nodes are the interpretable objects of networks. In contrast, flow-based community detection methods, such as Markov stability, provide a network-based viewpoint for dimensionality reduction, as they aim to identify groups of nodes that collectively affect the dynamics in the same way.

6.5 Further Discussions and References

This section has focused exclusively on Markov stability as an example of a flow-based framework for community detection. This choice was motivated by the flexibility of the framework and its clear connections with modularity and the concept of timescale separation in dynamical systems. However, other approaches have been proposed in the literature, often sharing the intuitive idea that a partition is good if random walkers remain confined for long times inside communities before escaping them (e.g., Piccardi, 2011). A popular alternative to Markov stability is the map equation (Rosvall & Bergstrom, 2008), which is centered around the idea that community structure allows us to compress the information required to describe the trajectory of a random walker on a network. While not part of the original map equation formalism, the map equation can be equipped with a timescale that determines the (sampling) rate at which one observes the random-walk trajectory (Schaub, Lambiotte, & Barahona, 2012), akin to Markov stability.

Recent works have also shown that spectral methods associated to the so-called non-backtracking random walks can successfully recover communities in sparse networks up to a well-defined theoretical (Krzakala et al., 2013) in certain random graph models. This result could, for example, be explored further within the framework developed in this section by constructing a Markov stability based on non-backtracking walks. Future research directions also include the possibility to design flow-based methods for temporal networks, where the random walkers would be diffusing on a network topology that evolves itself in time (Holme & Saramäki, 2019; Masuda & Lambiotte, 2020; Mucha et al., 2010), for hypergraphs, where interactions between nodes are not necessarily pairwise (Carletti, Fanelli, & Lambiotte, 2020; Eriksson et al., 2020), and for graphons, that can be seen as continuous generalisations of networks (Klimm, Jones, & Schaub, 2021).

7 Dynamical Methods for Disassortative Communities and General Block Structures

The previous section has shown that a dynamical perspective can help to design quality functions to detect assortative communities at different scales. In this section, we turn our attention to more general block structures. In the specific case of disassortative communities, following our discussion at the end of Section 3.3, a quality function for disassortative communties could be derived by extending the approach of Eq. (3.12), that is, by building a quality function based on the $C \times C$ clustered covariance matrix (6.5) and keeping the off-diagonal elements rather than the trace (i.e., minimise Markov stability, as discussed in Delvenne et al. (2013, 2010)). Here, rather than concentrating on the diffusion-based formalism of the previous section, we will discuss a related but different perspective, which has close connections to the active research area of network embeddings (Grover & Leskovec, 2016; Perozzi, Al-Rfou, & Skiena, 2014).

The purpose of embedding techniques is to map the nodes of a network into a low-dimensional metric vector-space, where proximity in the embedding is associated to node-similarity in the network. The advantage of such a procedure is that once we have obtained a node embedding, we can repurpose the large array of methods available to analyse vector space data for the study of networks by applying them to the embeddings of the nodes. In most cases such embedding techniques try to map nodes that are well connected in the network to similar embedding coordinates, and hence clustering the resulting points in the vector space can be used to uncover assortative communities. In the following we will first provide a brief survey of such (spectral) embedding procedures. We then discuss how we can use a dynamical perspective to define more generalised notions of node similarity and, as a result, can derive embeddings from which more general block structures can be extracted.

7.1 Kernels and Embeddings for Assortative Communities

Intuitively, a network (node) embedding is a map which assigns each node i in the node set $\mathcal{V}$ of a network $\mathcal{G}$ to a point in the Euclidean space $\mathbb{R}^{m_0}$, whose dimension m_0 is typically much smaller than the number of nodes n. Each node i is mapped to a m_0-dimensional vector $\boldsymbol{x}^{(i)}$, and the whole network is thus represented by a cloud of points in $\mathbb{R}^{m_0}$. An obvious application for node embeddings is graph drawing, but network embeddings are also central tools for clustering and node classification.

A classical method to derive a node embedding is the so-called Laplacian eigenmap (Belkin & Niyogi, 2001), whose goal is to find an embedding of a (connected, undirected) graph by minimising:

$$F(\boldsymbol{X}) = \frac{1}{2}\sum_{i,j=1}^{n} A_{ij}\|\boldsymbol{x}^{(i)} - \boldsymbol{x}^{(j)}\|^2 = \mathrm{Tr}\boldsymbol{X}^\top \boldsymbol{L}\boldsymbol{X}, \tag{7.1}$$

$$\text{s.t.} \quad \boldsymbol{X}^\top \boldsymbol{K}\boldsymbol{X} = \boldsymbol{I} \text{ and } \boldsymbol{X}^\top \boldsymbol{K}\mathbf{1} = \mathbf{0}. \tag{7.2}$$

where, as usual, $\boldsymbol{K}$ is the diagonal matrix of degrees and where the rows of the matrix $\boldsymbol{X} = [\boldsymbol{x}^{(1)}, \ldots, \boldsymbol{x}^{(n)}]^\top \in \mathbb{R}^{n\times m_0}$ contain the embedding coordinates. The embedding dimension m_0 is a free parameter and the constraints guarantee that the embedding does not collapse to a subspace of dimension smaller than m_0.

The attentive reader will have noticed that the above problem is very close to the minimal cut problem discussed in Section 3.2, and indeed the above objective function is small when nodes that are connected in the graph have an embedding with a small distance. Let $\boldsymbol{v}^{(i)}$ be the solutions of the generalised eigenvalue problem $\boldsymbol{L}\boldsymbol{v}^{(i)} = \lambda_i \boldsymbol{K}\boldsymbol{v}^{(i)}$, ordered according to increasing eigenvalues $\lambda_1 = 0 \leq \lambda_2 \leq \cdots \leq \lambda_n$. Following the same arguments as in Section 3.2, it is straightforward to show that the cost function is minimised by concatenating the m_0 generalised eigenvectors $\boldsymbol{v}^{(2)}, \ldots, \boldsymbol{v}^{(m_0+1)}$ of the Laplacian.

The above results showcase the intimate relations between graph embeddings and spectral methods for graph partitioning. If we want to use the embedding coordinates for the purpose of community detection, we are, however, not interested in the explicit embedding coordinates of the nodes, but in measures of similarity or distance, which can then be used to cluster the node embedding (and thus the network). Hence, graph kernel functions, as discussed in Section 2.5.2, which bypass the computation of explicit embeddings, are often considered instead of explicit embedding computations. Note, for instance, that the coordinates (2.28) associated to the heat kernel are closely related to the Laplacian eigenmaps $\boldsymbol{X}$ obtained from optimising Eq. (7.1).

As we have discussed extensively in the previous sections, the spectral properties of the Laplacian appearing in eigenmaps are also directly related to properties of random walks on networks. In fact, there exist a range of embedding techniques that depend even more explicitly on random walks, such as DeepWalk (Perozzi et al., 2014) and node2vec (Grover & Leskovec, 2016), where the underlying idea is to use trajectories of random walks over a fixed number of steps to characterise the neighbourhood of each node to be embedded. All these embedding techniques are based on the principle that 'proximity' of nodes in the network, as measured via a diffusion process, should lead to proximity in the vector space. For this reason, running standard clustering algorithms like k-means on the resulting data points in the embedding leads to communities that are assortative in nature (Tian et al., 2014). This idea also underpins the popular Walktrap algorithm (Pons & Latapy, 2005), where the so-called random-walk distance is calculated between the nodes, and then a standard agglomerative hierarchical clustering algorithm is used to uncover communities. Kernel methods, for instance, based on the heat kernel (2.27), also have intimate connections with community detection and block modelling (Kloumann, Ugander, & Kleinberg, 2017).

7.2 Dynamical Embeddings for General Linear Dynamics

The previous section discussed how network embeddings and kernels can be constructed from the properties of random-walk processes on networks. Here we present a framework introduced in Schaub (2014); Schaub et al. (2019a) that differs in three ways from these canonical methods.

First, we will be concerned with embeddings derived from general linear dynamics, which may approximate more appropriately dynamics of real-world systems, for example, in situations such as epidemic spreading that do not show a diffusive, but a multiplicative behaviour. Considering general linear dynamics also provides the flexibility to define embeddings on networks with signed edges, for which we cannot define a diffusion process in a simple way.

Second, the dynamical embeddings that we will consider here are defined for general directed networks, without the requirement that they are strongly connected. In contrast, for random-walk-based methods, we often need a strongly connected graph to ensure an ergodic dynamics. While this can be ensured by using the teleportation trick discussed above, this inadvertently corresponds to a perturbation of the random-walk dynamics, which may not be desired. Moreover, using the teleportation trick adds an extra parameter to the analysis, the teleportation probability.

Finally, for the more general dynamical framework considered here, the proximity of two nodes in the embedding will not depend on the proximity of nodes within the network. Rather, two nodes will be considered similar, and accordingly have similar embedding coordinates, if they have a similar effect on the overall network dynamics. As we will see, this dynamical viewpoint enables us to uncover more general block structures than assortative communities. At the same time, however, it turns out that this dynamical notion of similarity generalises the diffusion-based methods discussed before in certain conditions, and we can, for example, recover the continuous-time Markov stability framework for undirected network as a special case (Schaub et al., 2019a).

Note that for the remainder of this section, we will consider *directed networks* by default, unless otherwise stated. In order to derive a dynamical embedding of the nodes in the network, we build upon the general theory of linear systems. Specifically, our embedding associates to each node the trajectory of its (zero-state) impulse response. We can then use these impulse responses, parametrised by time t, as embeddings for any two nodes, and use the resulting coordinates, for example, to define a dynamical similarity measure between two nodes. To introduce these ideas, let us reconsider the case of a discrete-time random-walk dynamics from this perspective:

$$\boldsymbol{p}^{\top}(t+1) = \boldsymbol{p}^{\top}(t)\boldsymbol{T} \Leftrightarrow \boldsymbol{p}(t+1) = \boldsymbol{T}^{\top}\boldsymbol{p}(t). \tag{7.3}$$

Let us now assume that we inject an unit impulse to a given node i at time $t = 0$, such that $\boldsymbol{p}(0) = \boldsymbol{e}_i$ is the indicator vector with a value of 1 at position i and zero otherwise. The impulse response of the system at time t is now given by the vector $\boldsymbol{p}_i(t)$ defined by $\boldsymbol{p}_i(t) = [\boldsymbol{T}^{\top}]^t \boldsymbol{e}_i$, which is simply the ith row of the t-step transition matrix $\boldsymbol{T}^t$ of a discrete-time random walk, interpreted as column vector. In other words, we first assign all probability mass to node i at time $t = 0$ and then observe how it spreads throughout the network over time.

This procedure can be repeated for each node i, thus defining an embedding for each node via the map $i \mapsto \boldsymbol{p}_i(t)$ into the n-dimensional space of node signals. To capture the relation between two nodes, different choices of similarity functions between the vectors $\boldsymbol{p}_i(t)$ are possible. For simplicity, we consider here the kernel defined by the bilinear inner product

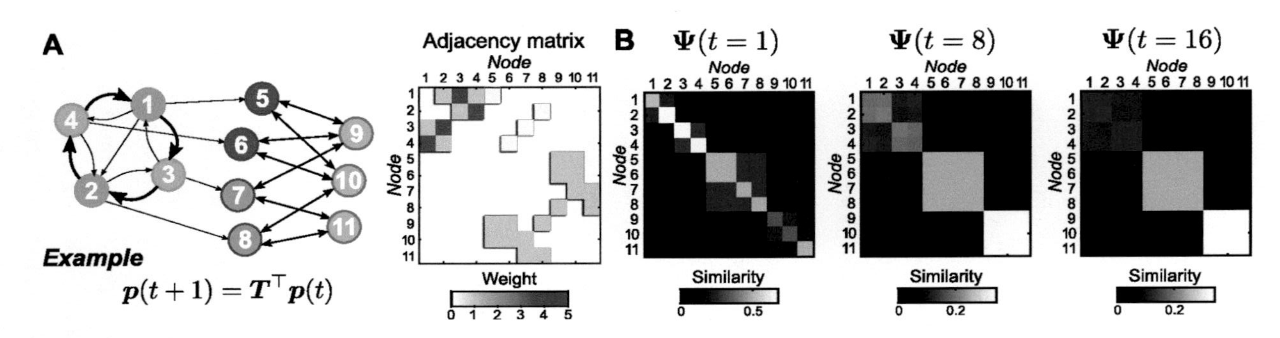

Figure 12 Dynamical similarity measures for random walks. A Visualisation of a (not strongly connected) directed network and its adjacency matrix. There are three main types of nodes identified by their different colours (subgroups within those three groups indicated by lighter colour). **B** The block structure in the similarity matrix $\mathbf{\Psi}(t) = \mathbf{T}^t[\mathbf{T}^t]^\top$, where $\mathcal{W}$ was chosen to be the identity matrix, identifies the dynamical role of the nodes at different times. Note that nodes within the cyan and violet groups are not connected to each other (i.e., the grouping is not assortative). Figure adapted and reproduced from Schaub et al. (2019a) with permission.

$$\Psi(t) = \left[\psi_{ij}(t)\right]_{i,j=1,\dots,n} \tag{7.4}$$
$$\text{with} \quad \psi_{ij}(t) = \boldsymbol{p}_i(t)^\top \mathcal{W} \boldsymbol{p}_j(t),$$

where the weighting matrix $\mathcal{W}$ allows, in general, to give more or less importance to certain nodes, for instance based on their degrees. This kernel is directly associated to a distance matrix $\boldsymbol{D}^{(2)}(t)$, whose entries correspond to a squared Euclidean distance of the form

$$D_{ij}^{(2)}(t) = \|\mathcal{W}^{\frac{1}{2}}\left(\boldsymbol{p}_i(t) - \boldsymbol{p}_j(t)\right)\|^2 = \psi_{ii} + \psi_{jj} - 2\psi_{ij}. \tag{7.5}$$

As in the case of Markov stability, $\Psi(t)$ and $\boldsymbol{D}^{(2)}(t)$ inherently depend on time, and the kernel and distance measures are thus expected to exhibit different patterns of similarity between nodes for different timescales in general.

Note that if the diffusion dynamics is ergodic and has a unique stationary state, $\lim_{t\to\infty} \boldsymbol{p}_i(t) = \boldsymbol{\pi}$ (which we do not assume, in general), then the distances between all nodes eventually become zero. As illustrated in Figure 12, the proximity induced by the dynamical embedding is, in contrast to the Markov stability framework, not associated to the presence of regions where the flow of probability is trapped. Instead, two nodes i, j are similar if they induce a similar state in the network at a particular timescale t. For the case of a discrete time diffusion this means that random walks starting on these two nodes tend to arrive at the same destination after time t. Accordingly, a high dynamical similarity does not necessitate direct proximity in the underlying graph.

This definition may also be understood in terms of a node equivalence that we may denote 'dynamical equivalence' (cf. Section 3.3 for a discussion on structural node equivalence notions). We provide an illustration of these concepts in Figure 12, where it can be seen that over particular timescales the kernel (7.4) naturally uncovers cyclic and bipartite structures that do not correspond to assortative blocks, and that are not directly apparent from inspection of the adjacency matrix.

These ideas can be directly extended from diffusion dynamics to any linear system. For instance, let us consider the case of the dual, continuous-time consensus dynamics $\dot{\boldsymbol{x}} = \boldsymbol{F}\boldsymbol{x} = -\boldsymbol{L}\boldsymbol{x}$. Applying an initial impulse on node i, $\boldsymbol{x}(0) = \boldsymbol{e}_i$, it is mapped onto the corresponding column of $\exp(t\boldsymbol{F})$ giving the response of the system at time t (i.e., $i \mapsto \boldsymbol{x}_i(t)$). Following the same arguments as before, the corresponding similarity matrix $\Psi(t)$ is defined as

$$\Psi(t) = \exp(t\boldsymbol{F})^\top \mathcal{W} \exp(t\boldsymbol{F}). \tag{7.6}$$

We can exploit this general vector space representation for various purposes. In the next section, we will illustrate how low-dimensional embeddings of the system can be constructed from the n-dimensional dynamical embedding, and discuss how to uncover dynamical modules in the system (i.e., groups of nodes that act approximately equivalent from a dynamical perspective).

7.3 Dimensionality Reduction and Detection of General Block Structures via Dynamical Similarities

The framework introduced in the previous section embeds nodes in a high-dimensional space, which can be of the same dimension as the number of nodes in the original graph. In this section, we outline how the above dynamical similarity measures may also be employed for dimensionality reduction. We start from the spectral decomposition of $\boldsymbol{\Psi}(t)$ into its eigenvectors $\boldsymbol{v}_1, \boldsymbol{v}_2, \ldots, \boldsymbol{v}_n$ with associated eigenvalues $\mu_1(t) \geq \mu_2(t) \geq \cdots \geq \mu_n(t)$.[16] We then define the mapping $i \mapsto \boldsymbol{\phi}_i(t)$,

$$\boldsymbol{\phi}_i(t) = \left[\sqrt{\mu_1}\,[\boldsymbol{v}_1]_i, \sqrt{\mu_2}\,[\boldsymbol{v}_2]_i, \ldots, \sqrt{\mu_n}\,[\boldsymbol{v}_n]_i\right]^\top, \tag{7.7}$$

which defines a signal in the spectral domain. Using simple algebraic manipulations, it can be verified that the squared dynamical distance measure (7.5) can be written as

$$D_{ij}^{(2)}(t) = \|\boldsymbol{\phi}_i(t) - \boldsymbol{\phi}_j(t)\|^2, \tag{7.8}$$

which shows that the distances in the two mappings are indeed the same. The advantage of adopting the spectral coordinates of $\boldsymbol{\Psi}$ is that this enables us to provide a principled approximation to the distance matrix. Specifically, a lower-dimensional embedding of the system minimising the error in the distance matrix $\boldsymbol{D}^{(2)}(t)$ can be obtained by keeping only the first c coordinates in each mapping $\boldsymbol{\phi}_i(t)$. Importantly, proximity in this low-dimensional embedding encodes the dynamical similarity between the nodes. Note that this procedure can be performed for any system matrix $\boldsymbol{F}$ and that it reduces to the so-called diffusion map embedding (Coifman et al., 2005; Lafon & Lee, 2006), when performed on the coordinates associated to the heat kernel in Eq. (2.28).

The similarity matrix $\boldsymbol{\Psi}(t)$, or equivalently the distance matrix $\boldsymbol{D}^{(2)}(t)$, can also help uncover general (dynamical) block structure in a network (cf. the example of Figure 12). For instance, this can be done by performing k-means clustering in the associated embedding. Alternatively, we may design quality functions based on $\boldsymbol{\Psi}(t)$, similar to Modularity and Markov stability, and define the quality of a partition in terms of the dynamical similarity of nodes inside

[16] Note that these are not the eigenvectors of $\boldsymbol{F}$, in general, and the eigenvectors $\boldsymbol{v}$ may not be constant in time; see Schaub et al. (2019a) for further discussion.

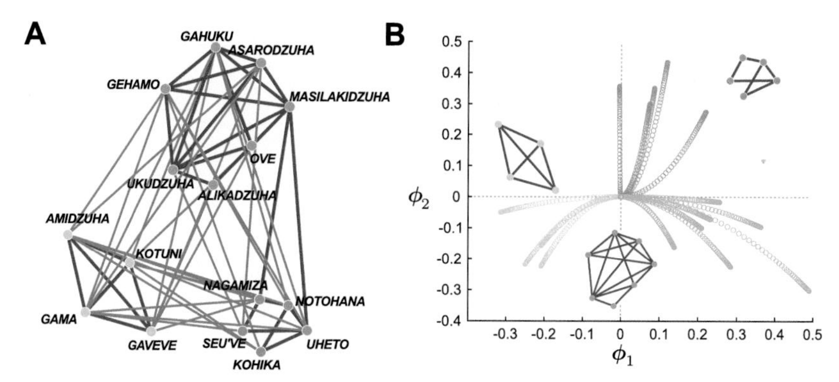

Figure 13 Dynamical similarity and embeddings signed networks. A We consider a network of 16 highland tribes with positive and negative interactions (see text). **B** The time-dependent dynamical similarity measure and associated embeddings enable us to identify the three main groups within this network. Figure adapted and reproduced from Schaub et al. (2019a) with permission.

clusters. Interestingly, similar results may also be achieved by changing the weighting matrix $\mathcal{W}$ to correspond to a projection matrix instead of a diagonal matrix. Since this has the effect of first projecting out a certain subspace from the state trajectories before computing a similarity score via an inner product, such an adjustment of the weighting matrix has a similar effect to choosing a certain null model, for example, within the context of Modularity. We refer to Schaub et al. (2019a) for a more extended discussion.

To showcase some of the facets of dynamical similarity, let us examine another illustrative example. Specifically, we consider the signed network of relationships between 16 tribal groups in New Guinea (Hage, Harary, & Harary, 1983; Read, 1954), as depicted in Figure 13A. The relationships between different tribes are either sympathetic (red edges) or antagonistic (blue edges). A simple model for consensus dynamics in such a signed network can be built on the principles that 'the friend of a friend is a friend' and 'the enemy of an enemy is a friend' (Altafini, 2012):

$$\dot{\boldsymbol{x}} = -\boldsymbol{L}_s \boldsymbol{x}, \tag{7.9}$$

where the state vector of the nodes is given by $\boldsymbol{x}$ and the signed Laplacian is defined as $\boldsymbol{L}_s = \boldsymbol{K}_s - \boldsymbol{A}_s$. Here $\boldsymbol{A}_s$ is the signed adjacency matrix of the network with positive and negative entries, and $\boldsymbol{K}_s$ is the diagonal matrix containing the weighted *absolute* node degrees, that is, $[\boldsymbol{K}_s]_{ii} = \sum_j |[\boldsymbol{A}_s]_{ij}|$ and zero otherwise. It can be shown that the signed Laplacian is positive semi-definite and that it reduces to the standard Laplacian for a network with non-negative weights (Altafini, 2013; Kunegis et al., 2010).

Based on the signed Laplacian dynamics discussed earlier, we can now define the following node similarity kernel:

$$\Psi(t) = \exp(-\boldsymbol{L}_s t)^\top \exp(-\boldsymbol{L}_s t), \tag{7.10}$$

with associated dynamical (spectral) embeddings $\boldsymbol{\phi}_i$. If we consider only the first two components of this embedding for each node i over time t, we obtain a trajectory for each node that describes how this (truncated) embedding evolves with time. This is depicted in Figure 13B. Based on this embedding, we can now partition the graph into groups by applying k-means clustering to the computed embeddings and obtain a split of the network into three dominant groups (as encoded in colour in Figure 13). As this example illustrates, the dynamical framework developed here can also be applied without modification to signed networks. For a more detailed discussion on this and other signed network examples, see Schaub et al. (2019a).

7.4 Node Equivalence Classes from Linear Dynamics

As discussed in Section 5, symmetries in the system matrix $\boldsymbol{F}$ of a dynamics $\dot{\boldsymbol{x}} = \boldsymbol{F}\boldsymbol{x}$ can lead to a dynamically equivalent behaviour of certain subsets of nodes within a network. Here we briefly sketch how we can recover such dynamical node roles by inspecting the impulse response dynamics of a linear system, in a way extending the method of the previous sections. To illustrate this idea, consider the graph shown in Figure 14A with an associated consensus dynamics $\dot{\boldsymbol{x}} = -\boldsymbol{L}\boldsymbol{x}$. Analysing the dynamical similarity measure $\boldsymbol{\Psi}(t)$ (see Figure 14B for the case $t = 0.1$) as discussed earlier reveals that there are three sets of nodes, as indicated by colour in Figure 14A, that have approximately the same dynamical effect on the system over a range of timescales.

Instead of taking the inner product between the impulse responses of each node as similarity measure $\boldsymbol{\Psi}$, let us now consider possible node permutations before taking the inner product. For example, we can define the similarity score $\theta_{ij}(t) = \max_{\boldsymbol{\Gamma}} \boldsymbol{x}_i(t)^\top \boldsymbol{\Gamma} \boldsymbol{x}_j(t)$, where $\boldsymbol{x}_i(t)$ corresponds to the impulse response of node i and we recall that $\boldsymbol{\Gamma}$ is a permutation matrix. When using this similarity score, we can find the grouping shown in Figure 14C. As can indeed be verified, the graph partition found in such a way groups nodes that have exactly the same influence on other nodes in the graph up to a permutation of the node labels (see Figure 14D). Importantly, such nodes may have a very different position in the graph and may not be in close proximity to each other.

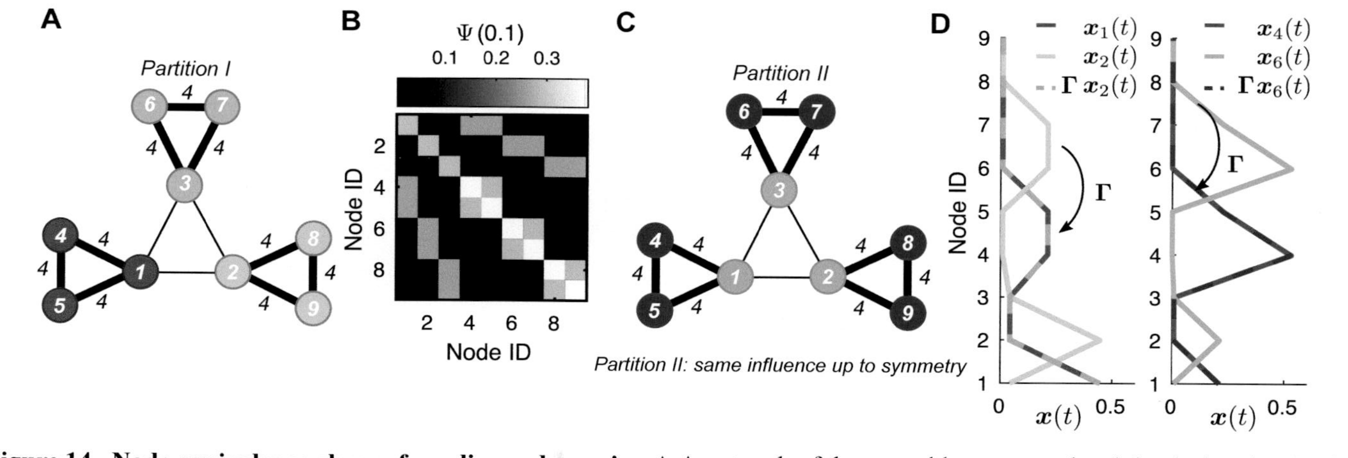

Figure 14 Node equivalence classes from linear dynamics. **A** A network of three weakly connected weighted triangles. **B** The corresponding dynamical similarity matrix $\mathbf{\Psi}(t)$, here shown for $t = 0.1$, shows that the graph has some clear dynamical community structure, in line with the colour coding displayed in (B). **C** An alternative partition of the network according to dynamical node roles, that is, sets of nodes that have the same dynamical impact on the graph modulo node relabelling, as demonstrated in **D**. Figure adapted and reproduced from Schaub et al. (2019a) with permission.

Following our discussions in Section 5, we remark upon another interesting duality between these two types of partitions. Namely, when finding dynamical graph partitions via an inspection of $\boldsymbol{\Psi}(t)$, we aim to group nodes that have the same effect on the (same nodes in the) network, over a particular timescale t. In contrast, we would expect nodes that serve the same role in the network, in the sense of the measure $\theta_{ij}(t)$, to have consistent influence over *all timescales* due to the symmetry properties of the system, which will affect the behaviour irrespective of the considered timescales (cf. Section 5). This may be exploited by computing a measure such as $\theta_{ij}(t)$ for multiple times of t and then finding the best possible permutation maximising all these scores jointly (Cason, 2014).

In order to avoid the (computationally expensive) search over the space of all permutation matrices, one can further adopt the following strategy: instead of using the measurements $\boldsymbol{x}_i(t)$ directly, we may for each time extract a feature from $\boldsymbol{x}_i(t)$ that is permutation invariant (i.e., some measurement function f for which $f(\boldsymbol{x}_i(t)) = f(\boldsymbol{\Gamma}\boldsymbol{x}_i(t))$ for any partition $\boldsymbol{\Gamma}$). An example for such a function f would be the mean, sum, max, min, or some kind of entropy function. Interestingly, it can be shown that this strategy effectively underpins a number of node role extractions (or node similarity measures) proposed in the literature (Blondel et al., 2004; Cason, 2014; Cooper & Barahona, 2010; Rossi et al., 2020). Moreover, similar strategies are also employed in the context of so-called graph neural networks (Wu et al., 2020), which have recently gained prominence for studying learning problems associated to graphs.

7.5 Further Discussions and References

In this section, we have exclusively considered Euclidian embeddings. In recent years, several works have shown the potential benefits of adopting a non-Euclidian perspective to embed networks. In particular, hyperbolic embeddings have the advantage of naturally representing networks with a broad degree distribution, as observed empirically (Serrano & Boguná, 2021). Recent methods for community detection based on hyperbolic embeddings include Faqeeh, Osat, and Radicchi (2018). This whole section has focused on linear dynamics, building on tools from linear response theory, but the possibility of defining the similarity between nodes from the response to an impulse could in principle be defined for non-linear dynamics,[17] and be a promising venue for future research.

[17] Note that in the case of non-linear dynamics, extra attention must be paid, as the amplitude of the impulse may in general lead to different responses.

8 Perspectives

In this Element we have tried to provide an overview of the interrelated aspects of community structure and linear dynamics on networks. In Sections 4 and 5, we have identified how assortative communities or more general block structures affect dynamical processes. As we discussed, such structures translate into specific properties for the eigenvalues (timescale separation) and eigenvectors (symmetries) of the system matrix governing the dynamics. In Sections 6 and 7, we considered the complementary viewpoint and considered how dynamics can enable the extraction of important information from a network and, more specifically, uncover block structures hidden in large networks. To this end, we have discussed several quality functions to uncover communities at different resolutions, depending on the timescale of the associated dynamics, and shown important connections with a range of network-theoretical concepts, including the Newman–Girvan modularity and embeddings techniques. We hope that this Element offers a coherent journey through the topic of dynamics and modularity on networks. However, it is important to emphasise that our account is but one specific path through a network of interrelated research topics. Although we tried to provide additional pointers to the literature in the 'Further Discussions and References' at the end of each section, we unfortunately had to leave out a wide range of interesting material to limit the scope of this Element. While we won't be able to provide an exhaustive overview here as well, in closing we would like to add some selected broader perspectives on what we believe are important themes for future research.

An important and active field of research focuses on how to incorporate **uncertainties** in network analysis. Indeed, as in any empirical dataset, networks that are collected from real-world data are incomplete and subject to measurement errors, and the statistical properties of these errors are often unknown. While the problem of measurement errors is well documented, especially for the analysis of social networks (Almquist, 2012; Borgatti, Carley, & Krackhardt, 2006; Holland & Leinhardt, 1973; Kossinets, 2006), it has regained interest in recent years in the context of network analysis (Martin, Ball, & Newman, 2016; Newman, 2018b; Ruggeri & De Bacco, 2019; Young, Cantwell, & Newman, 2020), motivated by the observation that most current techniques for network analysis implicitly assume, and are thus limited to, networks that are perfectly known. In Section 4.3.2, we touched on this point when estimating the impact of noise on the spectral properties of the system matrix, and thus on linear dynamics, but all our results on community and block structure detection, except for our short discussion on robustness in Section 6.2, did not consider noisy observations. In our view, future research is needed to properly capture the uncertainty in the measurement of networks and to clarify how uncertainty about the existence of edges translates into uncertainty in topological descriptors, including degree distribution (Stumpf, Wiuf, & May, 2005), centrality measures (Avella-Medina et al., 2020; Stamm et al., 2020; Wagner, Singer, & Karimi, 2017), or community structure (Peixoto, 2018; Smiljanić, Edler, & Rosvall, 2020).

We have exclusively focused on linear dynamics in this Element, which enabled us to exploit the correspondence between the timescale of a process and the eigenvalues of the corresponding system matrix. However, in many situations, linear dynamics is only an approximation of the **non-linear dynamics** in a system. The field of non-linear dynamics on networks is more fragmented, as many results depend on the details of the dynamical model. Different types of models are prevalent in different domains; for example, synchronisation (Arenas et al., 2008) versus epidemic spreading (Pastor-Satorras et al., 2015). For this reason, it is more difficult to provide a general overview such as the one we provided here for linear processes, even if some of the tools and phenomena that we discussed here for linear dynamics can be translated to a non-linear context; for example, externally equitable partitions (Schaub et al., 2016) and timescale separation (Arenas, Díaz-Guilera, & Pérez-Vicente, 2006) for non-linear models of synchronisation. Research has shown that non-linear dynamics on modular networks may lead to a rich set of dynamical phenomena. This includes the emergence of 'chimera states' in networks of oscillators, where the system splits into synchronised and desynchronised subpopulations (Abrams et al., 2008), or the stable coexistence of different opinions in social dynamics (Lambiotte, Ausloos, & Hołyst, 2007). In order to build a more general framework for non-linear dynamics on modular networks, some potential avenues for research are the study of network symmetries (Golubitsky & Stewart, 2006, 2015), or the use of Koopman operator theory (Mauroy, Susuki, & Mezić, 2020), in which the non-linear dynamics are lifted to a linear, but infinite-dimensional space.

We started the introduction of this Element by emphasising the power of the network paradigm to model complex systems. Despite its many successes, the abstraction of a system in terms of nodes and edges also has some fundamental modelling limitations that have become more apparent with the increasing availability of richer relational data recently. Within the emerging field of **higher-order networks**, researchers have tested the limits of the network paradigm and proposed extensions with generalised interactions, including multiplex networks (De Domenico et al., 2013), higher-order Markov models for networks (Lambiotte et al., 2019), and multiway networks (Barbarossa & Sardellitti, 2020; Battiston et al., 2020; Schaub et al., 2021). Multiplex networks aim at modelling systems where nodes can be connected by different types of interactions. Higher-order Markov models do not take edges as fundamental units of connectivity, but walks of length generally greater than one. Multiway networks also question the role of edges as fundamental units and focus instead on interactions involving more than two nodes, as in hypergraphs, for instance. In each type of higher-order model, the notion of connectivity is altered, hence leading to different types of diffusion and community structure. Despite its dynamism (see, e.g., Chodrow, Veldt, and Benson, 2021; Eriksson et al., 2021; Schaub et al., 2020), there are still many open questions in this field of research, calling for future investigations to properly comprehend the relations between modularity and dynamics on higher-order networks.

References

Abbe, E. (2017). Community detection and stochastic block models: recent developments. *The Journal of Machine Learning Research, 18*(1), 6446–6531.

Abrams, D. M., Mirollo, R., Strogatz, S. H., & Wiley, D. A. (2008). Solvable model for chimera states of coupled oscillators. *Physical Review Letters, 101*(8), 084103.

Ahn, Y.-Y., Bagrow, J. P., & Lehmann, S. (2010). Link communities reveal multiscale complexity in networks. *Nature, 466*(7307), 761–764.

Almquist, Z. W. (2012, October). Random errors in egocentric networks. *Social Networks, 34*(4), 493–505. doi: https://doi.org/10.1016/j.socnet.2012.03.002

Alpert, C. J., & Kahng, A. B. (1995). Recent directions in netlist partitioning: a survey. *Integration, 19*(1–2), 1–81.

Altafini, C. (2012). Dynamics of opinion forming in structurally balanced social networks. *PloS One, 7*(6), e38135.

Altafini, C. (2013, April). Consensus problems on networks with antagonistic interactions. *IEEE Transactions on Automatic Control, 58*(4), 935–946. doi: https://doi.org/10.1109/TAC.2012.2224251

Arenas, A., Díaz-Guilera, A., Kurths, J., Moreno, Y., & Zhou, C. (2008). Synchronization in complex networks. *Physics Reports, 469*(3), 93–153.

Arenas, A., Díaz-Guilera, A., & Pérez-Vicente, C. J. (2006). Synchronization reveals topological scales in complex networks. *Physical Review Letters, 96*(11), 114102.

Asllani, M., Lambiotte, R., & Carletti, T. (2018). Structure and dynamical behavior of non-normal networks. *Science Advances, 4*(12), eaau9403.

Avella-Medina, M., Parise, F., Schaub, M. T., & Segarra, S. (2020). Centrality measures for graphons: accounting for uncertainty in networks. *IEEE Transactions on Network Science and Engineering, 7*(1), 520–537. doi: https://doi.org/10.1109/TNSE.2018.2884235

Aynaud, T., Blondel, V. D., Guillaume, J. L., & Lambiotte, R. (2013). Multilevel local optimization of modularity. In: Bichot, C.-E., Siarry, P. (eds.), *Graph Partitioning*, (pp. 315–345). John Wiley and Sons, Hoboken, NJ.

Banisch, R., & Conrad, N. D. (2015). Cycle-flow–based module detection in directed recurrence networks. *EPL (Europhysics Letters), 108*(6), 68008.

Barabási, A.-L., et al. (2016). *Network science*. Cambridge University Press, Cambridge, UK.

Barbarossa, S., & Sardellitti, S. (2020). Topological signal processing: making sense of data building on multiway relations. *IEEE Signal Processing Magazine*, *37*(6), 174–183.

Battiston, F., Cencetti, G., Iacopini, I. et al. (2020). Networks beyond pairwise interactions: structure and dynamics. *Physics Reports*, *874*, 1–92.

Belkin, M., & Niyogi, P. (2001). Laplacian eigenmaps and spectral techniques for embedding and clustering. *Advances in Neural Information Processing Systems*, *14*, 585–591.

Bhatia, R. (2013). *Matrix analysis* (vol. 169). Springer Science & Business Media, London.

Blondel, V. D., Gajardo, A., Heymans, M., Senellart, P., & Van Dooren, P. (2004). A measure of similarity between graph vertices: applications to synonym extraction and web searching. *SIAM Review*, *46*(4), 647–666.

Blondel, V. D., Guillaume, J.-L., Lambiotte, R., & Lefebvre, E. (2008). Fast unfolding of communities in large networks. *Journal of Statistical Mechanics: Theory and Experiment*, *2008*(10), P10008.

Bohlin, L., Edler, D., Lancichinetti, A., & Rosvall, M. (2014). Community detection and visualization of networks with the map equation framework. In: Ding, Y., Rousseau, R., & Wolfram, D. (eds.), *Measuring scholarly impact*, (pp. 3–34). Springer, New York.

Borgatti, S. P., Carley, K. M., & Krackhardt, D. (2006). On the robustness of centrality measures under conditions of imperfect data. *Social Networks*, *28*(2), 124–136. doi: https://doi.org/10.1016/j.socnet.2005.05.001

Brandes, U. (2005). *Network analysis: methodological foundations*, (vol. 3418). Springer Science & Business Media, London.

Brandes, U., Delling, D., Gaertler, M. et al. (2007). On modularity clustering. *IEEE Transactions on Knowledge and Data Engineering*, *20*(2), 172–188.

Brin, S., & Page, L. (1998). The anatomy of a large-scale hypertextual web search engine. *Computer Networks and ISDN Systems*, *30*(1–7), 107–117.

Broido, A. D., & Clauset, A. (2019). Scale-free networks are rare. *Nature Communications*, *10*(1), 1–10.

Bui-Xuan, B.-M., & Jones, N. S. (2014). How modular structure can simplify tasks on networks: parameterizing graph optimization by fast local community detection. *Proceedings of the Royal Society A: Mathematical, Physical and Engineering Sciences*, *470*(2170), 20140224.

Bullo, F. (2019). *Lectures on network systems*. Kindle Direct Publishing. ISBN 978-1-986425-64-3.

Burt, R. S. (2004). Structural holes and good ideas. *American Journal of Sociology*, *110*(2), 349–399.

Cardoso, D. M., Delorme, C., & Rama, P. (2007). Laplacian eigenvectors and eigenvalues and almost equitable partitions. *European Journal of Combinatorics*, *28*(3), 665–673.

Carletti, T., Fanelli, D., & Lambiotte, R. (2020). Random walks and community detection in hypergraphs. *arXiv preprint arXiv:2010.14355*.

Cason, T. P. (2014). Role extraction in networks. (unpublished doctoral dissertation). Catholic University of Louvain.

Cavallari, S., Zheng, V. W., Cai, H., Chang, K. C.-C., & Cambria, E. (2017). Learning community embedding with community detection and node embedding on graphs. In *Proceedings of the 2017 ACM on Conference on Information and Knowledge Management*, (pp. 377–386).

Chan, A., & Godsil, C. D. (1997). Symmetry and eigenvectors. In: Hahn, G., & Sabidussi, G. (eds.), *Graph Symmetry*, (pp. 75–106). Springer, New York.

Chandra, A. K., Raghavan, P., Ruzzo, W. L., Smolensky, R., & Tiwari, P. (1996). The electrical resistance of a graph captures its commute and cover times. *Computational Complexity*, *6*(4), 312–340.

Chodrow, P. S., Veldt, N., & Benson, A. R. (2021). Hypergraph clustering: from blockmodels to modularity. *arXiv preprint arXiv:2101.09611*.

Chung, F., & Lu, L. (2002). Connected components in random graphs with given expected degree sequences. *Annals of Combinatorics*, *6*(2), 125–145.

Chung, F. R. (1997). *Spectral graph theory*, (vol. 92). American Mathematical Society, Providence, RI.

Coifman, R. R., Lafon, S., Lee, A. B. et al. (2005). Geometric diffusions as a tool for harmonic analysis and structure definition of data: diffusion maps. *Proceedings of the National Academy of Sciences of the United States of America*, *102*(21), 7426–7431.

Conrad, N. D., Weber, M., & Schütte, C. (2016). Finding dominant structures of nonreversible Markov processes. *Multiscale Modeling & Simulation*, *14*(4), 1319–1340.

Cooper, K., & Barahona, M. (2010). Role-based similarity in directed networks. *arXiv preprint arXiv:1012.2726*.

Dasgupta, A., Hopcroft, J. E., & McSherry, F. (2004). Spectral analysis of random graphs with skewed degree distributions. In *45th Annual IEEE Symposium on Foundations of Computer Science*, (pp. 602–610). doi: https://doi.org/10.1109/FOCS.2004.61.

De Domenico, M., Solé-Ribalta, A., Cozzo, E. et al. (2013). Mathematical formulation of multilayer networks. *Physical Review X*, *3*(4), 041022.

Delvenne, J.-C., & Libert, A.-S. (2011). Centrality measures and thermodynamic formalism for complex networks. *Physical Review E*, *83*(4), 046117.

Delvenne, J.-C., Schaub, M. T., Yaliraki, S. N., & Barahona, M. (2013). The stability of a graph partition: a dynamics-based framework for community detection. In A. Mukherjee, M. Choudhury, F. Peruani, N. Ganguly, & B. Mitra (eds.), *Dynamics On and Of Complex Networks, Volume 2* (pp. 221–242). Springer, New York. doi: https://doi.org/10.1007/978-1-4614-6729-8_11

Delvenne, J.-C., Yaliraki, S. N., & Barahona, M. (2010). Stability of graph communities across time scales. *Proceedings of the National Academy of Sciences*, *107*(29), 12755–12760. doi: https://doi.org/10.1073/pnas.0903215107

Derrida, B., & Flyvbjerg, H. (1986). Multivalley structure in Kauffman's model: analogy with spin glasses. *Journal of Physics A: Mathematical and General*, *19*(16), L1003.

Devriendt, K. (2020). Effective resistance is more than distance: Laplacians, simplices and the Schur complement. *arXiv preprint arXiv:2010.04521*.

Doreian, P., Batagelj, V., & Ferligoj, A. (2020). *Advances in Network Clustering and Blockmodeling*. John Wiley & Hoboken, NJ.

Egerstedt, M., Martini, S., Cao, M., Camlibel, K., & Bicchi, A. (2012). Interacting with networks: how does structure relate to controllability in single-leader, consensus networks? *Control Systems, IEEE*, *32*(4), 66–73. doi: https://doi.org/10.1109/MCS.2012.2195411

Eriksson, A., Edler, D., Rojas, A., & Rosvall, M. (2020). Mapping flows on hypergraphs. *arXiv preprint arXiv:2101.00656*.

Eriksson, A., Edler, D., Rojas, A., & Rosvall, M. (2021). Mapping flows on hypergraphs. *arXiv preprint arXiv:2101.00656*.

Estrada, E., & Hatano, N. (2008). Communicability in complex networks. *Physical Review E*, *77*(3), 036111.

Everett, M. G., & Borgatti, S. P. (1994). Regular equivalence: general theory. *Journal of Mathematical Sociology*, *19*(1), 29–52.

Expert, P., Evans, T. S., Blondel, V. D., & Lambiotte, R. (2011). Uncovering space-independent communities in spatial networks. *Proceedings of the National Academy of Sciences*, *108*(19), 7663–7668.

Faccin, M., Schaub, M. T., & Delvenne, J.-C. (2018). Entrograms and coarse graining of dynamics on complex networks. *Journal of Complex Networks*, *6*(5), 661–678.

Faccin, M., Schaub, M. T., & Delvenne, J.-C. (2020). State aggregations in Markov chains and block models of networks. *arXiv preprint arXiv:2005.00337*.

Faqeeh, A., Osat, S., & Radicchi, F.(2018).Characterizing the analogy between hyperbolic embedding and community structure of complex networks. *Physical Review Letters*, *121*(9), 098301.

Fiedler, M. (1973). Algebraic connectivity of graphs. *Czechoslovak Mathematical Journal*, *23*(2), 298–305.

Fiedler, M. (2011). *Matrices and Graphs in Geometry*. Cambridge University Press, Cambridge, UK. doi: https://doi.org/10.1017/cbo9780511973611

Fortunato, S. (2010). Community detection in graphs. *Physics Reports*, *486* (3–5), 75–174. doi: https://doi.org/10.1016/j.physrep.2009.11.002

Fortunato, S., & Barthelemy, M. (2007). Resolution limit in community detection. *Proceedings of the National Academy of Sciences*, *104*(1), 36–41.

Fortunato, S., & Hric, D. (2016). Community detection in networks: a user guide. *Physics Reports*, *659*, 1–44.

Fosdick, B. K., Larremore, D. B., Nishimura, J., & Ugander, J. (2018). Configuring random graph models with fixed degree sequences. *SIAM Review*, *60*(2), 315–355.

Fouss, F., Saerens, M., & Shimbo, M. (2016). *Algorithms and models for network data and link analysis*. Cambridge University Press, Cambridge, UK.

Ghasemian, A., Hosseinmardi, H., & Clauset, A. (2019). Evaluating overfit and underfit in models of network community structure. *IEEE Transactions on Knowledge and Data Engineering*, *32*(9), 1722–1735. doi: https://doi.org/10.1109/TKDE.2019.2911585.

Gleich, D. F. (2015). Pagerank beyond the web. *SIAM Review*, *57*(3), 321–363.

Godsil, C., & Royle, G. F. (2013). *Algebraic graph theory* (vol. 207). Springer Science & Business Media, London.

Golub, G., & Van Loan, C. (2013). *Matrix computations*. 4th ed. Johns Hopkins. University Press, Baltimore, MD.

Golubitsky, M., & Stewart, I. (2006). Nonlinear dynamics of networks: the groupoid formalism. *Bulletin of the American Mathematical Society*, *43*(3), 305–364.

Golubitsky, M., & Stewart, I. (2015). Recent advances in symmetric and network dynamics. *Chaos: An Interdisciplinary Journal of Nonlinear Science*, *25*(9), 097612.

Good, B. H., De Montjoye, Y.-A., & Clauset, A. (2010). Performance of modularity maximization in practical contexts. *Physical Review E*, *81*(4), 046106.

Grohe, M., Kersting, K., Mladenov, M., & Selman, E. (2014). Dimension reduction via colour refinement. In *European Symposium on Algorithms*, (pp. 505–516). Springer, Berlin, Heidelberg. https://doi.org/10.1007/978-3-662-44777-2_42

Grohe, M., & Schweitzer, P. (2020). The graph isomorphism problem. *Communications of the ACM*, *63*(11), 128–134.

Grover, A., & Leskovec, J. (2016). node2vec: Scalable feature learning for networks. In *Proceedings of the 22nd ACM SIGKDD International Conference on Knowledge Discovery and Data Mining*, (pp. 855–864).

Gvishiani, A. D., & Gurvich, V. A. (1987). Metric and ultrametric spaces of resistances. *Uspekhi Matematicheskikh Nauk*, *42*(6), 187–188.

Hage, P., Harary, F., & Harary, F. (1983). *Structural models in anthropology*. Cambridge Studies in Social Anthropology. No. 46. Cambridge University Press, Cambridge, UK.

Higham, N. J. (2008). *Functions of matrices: theory and computation*. SIAM.

Holland, P. W., Laskey, K. B., & Leinhardt, S. (1983). Stochastic blockmodels: first steps. *Social Networks*, *5*(2), 109–137.

Holland, P. W., & Leinhardt, S. (1973). The structural implications of measurement error in sociometry. *Journal of Mathematical Sociology*, *3*(1), 85–111.

Holme, P., & Saramäki, J. (2019). *Temporal Network Theory*. Springer.

Karrer, B., & Newman, M. E. J. (2011). Stochastic blockmodels and community structure in networks. *Physical Review E*, *83*(1), 016107. doi: https://doi.org/10.1103/PhysRevE.83.016107

Kivela, M., Arenas, A., Barthelemy, M. et al. (2014). Multilayer networks. *Journal of Complex Networks*, *2*(3), 203–271. doi: `https://doi.org/10.1093/comnet/cnu016`

Klein, D. J., & Randić, M. (1993). Resistance distance. *Journal of Mathematical Chemistry*, *12*(1), 81–95. doi: `https://doi.org/10.1007%2Fbf01164627`

Klimm, F., Jones, N. S., & Schaub, M. T. (2021). Modularity maximisation for graphons. *arXiv preprint arXiv:2101.00503*.

Kloumann, I. M., Ugander, J., & Kleinberg, J. (2017). Block models and personalized pagerank. *Proceedings of the National Academy of Sciences*, *114*(1), 33–38.

Komarek, A., Pavlik, J., & Sobeslav, V. (2015). Network visualization survey. In *Computational Collective Intelligence*, (pp. 275–284). Springer.

Kondor, R., & Lafferty, J. (2002). Diffusion kernels on graphs and other discrete input spaces. In *Proceedings of the ICML'02: Nineteenth International Joint Conference on Machine Learning*, (pp. 315–322).

Kossinets, G. (2006). Effects of missing data in social networks. *Social Networks*, *28*(3), 247–268. Accessed 1 October 2020 from `https://linkinghub.elsevier.com/retrieve/pii/S0378873305000511` doi: https://doi.org/10.1016/j.socnet.2005.07.002

Krzakala, F., Moore, C., Mossel, E. et al. (2013). Spectral redemption in clustering sparse networks. *Proceedings of the National Academy of Sciences*, *110*(52), 20935–20940.

Kunegis, J., Schmidt, S., Lommatzsch, A. et al. (2010). Spectral analysis of signed graphs for clustering, prediction and visualization. In *Proceedings of the 2010 SIAM International Conference on Data Mining (SDM)* (vol. 10, pp. 559–570). Society for Industrial and Applied Mathematics.

Lafon, S., & Lee, A. (2006). Diffusion maps and coarse-graining: a unified framework for dimensionality reduction, graph partitioning, and data set parameterization. *Pattern Analysis and Machine Intelligence, IEEE Transactions on*, *28*(9), 1393–1403. doi: https://doi.org/10.1109/TPAMI.2006.184

Lambiotte, R., Ausloos, M., & Hołyst, J. (2007). Majority model on a network with communities. *Physical Review E*, *75*(3), 030101.

Lambiotte, R., Delvenne, J.-C., & Barahona, M. (2014). Random walks, Markov processes and the multiscale modular organization of complex networks. *IEEE Transactions on Network Science and Engineering*, *1*(2), 76–90. doi: https://doi.org/10.1109/TNSE.2015.2391998

Lambiotte, R., & Rosvall, M. (2012). Ranking and clustering of nodes in networks with smart teleportation. *Physical Review E*, *85*(5), 056107.

Lambiotte, R., Rosvall, M., & Scholtes, I. (2019). From networks to optimal higher-order models of complex systems. *Nature Physics*, *15*(4), 313–320.

Langville, A. N., & Meyer, C. D. (2011). *Google's PageRank and beyond: the science of search engine rankings*. Princeton University Press.

Le, C. M., Levina, E., & Vershynin, R. (2017). Concentration and regularization of random graphs. *Random Structures & Algorithms*, *51*(3), 538–561.

Lei, J., & Rinaldo, A. (2015). Consistency of spectral clustering in stochastic block models. *The Annals of Statistics*, *43*(1), 215–237.

Leskovec, J., Lang, K. J., Dasgupta, A., & Mahoney, M. W. (2008). Statistical properties of community structure in large social and information networks. In *Proceedings of the 17th International Conference on World Wide Web* (pp. 695–704).

Lorrain, F., & White, H. C. (1971). Structural equivalence of individuals in social networks. *The Journal of Mathematical Sociology*, *1*(1), 49–80.

Malliaros, F. D., & Vazirgiannis, M. (2013). Clustering and community detection in directed networks: a survey. *Physics Reports*, *533*(4), 95–142.

Martin, T., Ball, B., & Newman, M. E. J. (2016). Structural inference for uncertain networks. *Physical Review E*, *93*(1), 012306.

Masuda, N., & Lambiotte, R. (2020). *A guide To Temporal Networks* (vol. 6). World Scientific.

Masuda, N., Porter, M. A., & Lambiotte, R. (2017). Random walks and diffusion on networks. *Physics Reports*, *716*, 1–58.

Mauroy, A., Susuki, Y., & Mezić, I. (2020). *The Koopman Operator in Systems and Control*. Springer.

McPherson, M., Smith-Lovin, L., & Cook, J. M. (2001). Birds of a feather: homophily in social networks. *Annual Review of Sociology*, *27*(1), 415–444.

Meilă, M. (2007). Comparing clusterings – an information based distance. *Journal of Multivariate Analysis*, *98*(5), 873–895.

Menczer, F., Fortunato, S., & Davis, C. A. (2020). *A First Course in Network Science*. Cambridge University Press.

Meunier, D., Lambiotte, R., & Bullmore, E. T. (2010). Modular and hierarchically modular organization of brain networks. *Frontiers in Neuroscience*, *4*, 200.

Milo, R., Shen-Orr, S., Itzkovitz, S. et al. (2002). Network motifs: simple building blocks of complex networks. *Science*, *298*(5594), 824–827.

Mucha, P. J., Richardson, T., Macon, K., Porter, M. A., & Onnela, J.-P. (2010). Community structure in time-dependent, multiscale, and multiplex networks. *Science*, *328*(5980), 876–878.

Newman, M. E., et al. (2003). Random graphs as models of networks. *Handbook of Graphs and Networks*, *1*, 35–68.

Newman, M. E. J. (2013). Spectral methods for community detection and graph partitioning. *Physical Review E*, *88*(4), 042822.

Newman, M. E. J. (2016). Community detection in networks: modularity optimization and maximum likelihood are equivalent. *arXiv preprint arXiv:1606.02319*.

Newman, M. E. J. (2018a). *Network*. Oxford University Press.

Newman, M. E. J. (2018b). Network structure from rich but noisy data. *Nature Physics*, *14*, 5.

Newman, M. E. J., & Girvan, M. (2004). Finding and evaluating community structure in networks. *Physical Review E*, *69*(2), 026113.

Nicosia, V., Mangioni, G., Carchiolo, V., & Malgeri, M. (2009). Extending the definition of modularity to directed graphs with overlapping communities. *Journal of Statistical Mechanics: Theory and Experiment*, *2009*(03), P03024.

O'Clery, N., Yuan, Y., Stan, G.-B., & Barahona, M. (2013). Observability and coarse graining of consensus dynamics through the external equitable partition. *Physical Review E*, *88*(4). doi: https://doi.org/10.1103/physreve.88.042805

Park, J., & Newman, M. E. (2004). Statistical mechanics of networks. *Physical Review E*, *70*(6), 066117.

Pastor-Satorras, R., Castellano, C., Van Mieghem, P., & Vespignani, A. (2015). Epidemic processes in complex networks. *Reviews of Modern Physics*, *87*(3), 925.

Pecora, L. M., Sorrentino, F., Hagerstrom, A. M., Murphy, T. E., & Roy, R. (2014). Cluster synchronization and isolated desynchronization in complex networks with symmetries. *Nature Communications*, *5*, 4079.

Peel, L., Larremore, D. B., & Clauset, A. (2017). The ground truth about metadata and community detection in networks. *Science Advances*, *3*(5), e1602548.

Peixoto, T. P. (2018). Reconstructing networks with unknown and heterogeneous errors. *Physical Review X*, *8*(4), 041011.

Peixoto, T. P. (2019). Bayesian stochastic blockmodeling. *Advances in Network Clustering and Blockmodeling*, 289–332.

Perozzi, B., Al-Rfou, R., & Skiena, S. (2014). Deepwalk: online learning of social representations. In *Proceedings of the 20th ACM SIGKDD International Conference on Knowledge Discovery and Data Mining* (pp. 701–710).

Piccardi, C. (2011). Finding and testing network communities by lumped Markov chains. *PLoS One*, *6*(11), e27028. doi: 10.1371/journal.pone.0027028

Pons, P., & Latapy, M. (2005). Computing communities in large networks using random walks. In: Yolum, P., Güngör, T., Gürgen, F., & Özturan, C. (eds.), *International symposium on computer and information sciences*, (pp. 284–293). Springer, Berlin, Heidelberg. doi: https://doi.org/10.1007/11569596_31

Porter, M., Onnela, J., & Mucha, P. (2009). Communities in networks. *Notices of the AMS*, *56*(9), 1082–1097, 1164–1166.

Porter, M. A., & Gleeson, J. P. (2016). Dynamical systems on networks. *Frontiers in Applied Dynamical Systems: Reviews and Tutorials*, *4*. doi: https://doi.org/10.1101/2021.01.21.427609

Proskurnikov, A. V., & Tempo, R. (2017). A tutorial on modeling and analysis of dynamic social networks. Part i. *Annual Reviews in Control*, *43*, 65–79. doi: https://doi.org/10.1016/j.arcontrol.2017.03.002

Putra, P., Thompson, T. B., & Goriely, A. (2021). Braiding braak and braak: staging patterns and model selection in network neurodegeneration. *bioRxiv*. Accessed at www.biorxiv.org.

Read, K. E. (1954). Cultures of the Central Highlands, New Guinea. *Southwestern Journal of Anthropology*, *10*(1), 1–43.

Reichardt, J., & Bornholdt, S. (2006). Statistical mechanics of community detection. *Physical Review E*, *74*(1), 016110.

Reichardt, J., & White, D. R. (2007). Role models for complex networks. *The European Physical Journal B*, *60*(2), 217–224.

Rényi, A., et al. (1961). On measures of entropy and information. In *Proceedings of the Fourth Berkeley Symposium on Mathematical Statistics and Probability Volume 1: contributions to the theory of statistics*.

Rohe, K., Chatterjee, S., Yu, B., et al. (2011). Spectral clustering and the high-dimensional stochastic blockmodel. *The Annals of Statistics*, *39*(4), 1878–1915.

Rombach, M. P., Porter, M. A., Fowler, J. H., & Mucha, P. J. (2014). Core-periphery structure in networks. *SIAM Journal on Applied Mathematics*, *74*(1), 167–190.

Rossetti, G., & Cazabet, R. (2018). Community discovery in dynamic networks: a survey. *ACM Computing Surveys (CSUR)*, *51*(2), 1–37.

Rossi, R. A., Jin, D., Kim, S. et al.. (2020). On proximity and structural role-based embeddings in networks: misconceptions, techniques, and applications. *ACM Transactions on Knowledge Discovery from Data (TKDD)*, *14*(5), 1–37.

Rosvall, M., & Bergstrom, C. T. (2008). Maps of random walks on complex networks reveal community structure. *Proceedings of the National Academy of Sciences*, *105*(4), 1118–1123.

Rosvall, M., Esquivel, A. V., Lancichinetti, A., West, J. D., & Lambiotte, R. (2014). Memory in network flows and its effects on spreading dynamics and community detection. *Nature Communications*, *5*, 4630. doi: https://doi.org/10.1038/ncomms5630

Ruggeri, N., & De Bacco, C. (2019, August). Sampling on networks: estimating eigenvector centrality on incomplete graphs. *arXiv:1908.00388v1 [cs.SI]*.

Salnikov, V., Schaub, M. T., & Lambiotte, R. (2016). Using higher-order Markov models to reveal flow-based communities in networks. *Scientific Reports*, *6*, 23194. doi: https://doi.org/10.1038/srep23194

Sanchez-Garcia, R. J. (2018). Exploiting symmetry in network analysis. *arXiv preprint arXiv:1803.06915*.

Schaeffer, S. E. (2007). Graph clustering. *Computer Science Review*, *1*(1), 27–64. doi: https://doi.org/10.1016/j.cosrev.2007.05.001

Schaub, M. T. (2014). Unraveling complex networks under the prism of dynamical processes: relations between structure and dynamics. (Doctoral dissertation). Imperial College London.

Schaub, M. T., Benson, A. R., Horn, P., Lippner, G., & Jadbabaie, A. (2020). Random walks on simplicial complexes and the normalized Hodge 1-Laplacian. *SIAM Review*, *62*(2), 353–391.

Schaub, M. T., Billeh, Y. N., Anastassiou, C. A., Koch, C., & Barahona, M. (2015). Emergence of slow-switching assemblies in structured neuronal networks. *PLOS Computational Biology*, *11*(7), e1004196.

Schaub, M. T., Delvenne, J.-C., Lambiotte, R., & Barahona, M. (2019a). Multiscale dynamical embeddings of complex networks. *Physical Review E*, *99*(6), 062308. doi: https://doi.org/10.1103/PhysRevE.99.062308

Schaub, M. T., Delvenne, J.-C., Lambiotte, R., & Barahona, M. (2019b). Structured networks and coarse-grained descriptions: a dynamical perspective. *Advances in Network Clustering and Blockmodeling*, pp. 333–361.

Schaub, M. T., Delvenne, J.-C., Rosvall, M., & Lambiotte, R. (2017, 02). The many facets of community detection in complex networks. *Applied Network Science*, *2*(1), 4. doi: 10.1007/s41109-017-0023-6

Schaub, M. T., Delvenne, J.-C., Yaliraki, S. N., & Barahona, M. (2012a). Markov dynamics as a zooming lens for multiscale community detection: non clique-like communities and the field-of-view limit. *PloS One*, *7*(2), e32210.

Schaub, M. T., Lambiotte, R., & Barahona, M. (2012b). Encoding dynamics for multiscale community detection: Markov time sweeping for the map equation. *Physical Review E*, *86*(2), 026112.

Schaub, M. T., O'Clery, N., Billeh, Y. N. et al. (2016). Graph partitions and cluster synchronization in networks of oscillators. *Chaos: An Interdisciplinary Journal of Nonlinear Science*, *26*(9), 094821.

Schaub, M. T., & Peel, L. (2020). Hierarchical community structure in networks. *arXiv preprint arXiv:2009.07196*.

Schaub, M. T., Zhu, Y., Seby, J.-B., Roddenberry, T. M., & Segarra, S. (2021). Signal processing on higher-order networks: Livin' on the edge... and beyond. *Signal Processing*, *187*, 108149. doi: https://doi.org/10.1016/j.sigpro.2021.108149

Serrano, M. A., & Boguná, M. (2021). *The shortest path to network geometry*. Cambridge University Press.

Serrano, M. A., Krioukov, D., & Boguná, M. (2008). Self-similarity of complex networks and hidden metric spaces. *Physical Review Letters*, *100*(7), 078701.

Shi, J., & Malik, J. (1997). Normalized cuts and image segmentation. In *Proceedings of IEEE Computer Society Conference on Computer Vision and Pattern Recognition* (pp. 731–737).

Shuman, D. I., Narang, S. K., Frossard, P., Ortega, A., & Vandergheynst, P. (2013). The emerging field of signal processing on graphs: extending high-dimensional data analysis to networks and other irregular domains. *IEEE Signal Processing Magazine*, *30*(3), 83–98.

Simon, H. A. (1962). The architecture of complexity. *Proceedings of the American Philosophical Society*, *106*(6), 467–482.

Simon, H. A., & Ando, A. (1961). Aggregation of variables in dynamic systems. *Econometrica: Journal of the Econometric Society*, 111–138.

Simpson, E. H. (1949). Measurement of diversity. *Nature*, *163*(4148), 688.

Smiljanić, J., Edler, D., & Rosvall, M. (2020). Mapping flows on sparse networks with missing links. *Physical Review E*, *102*(1), 012302.

Spielman, D. A., & Teng, S.-H. (2011). Spectral sparsification of graphs. *SIAM Journal on Computing*, *40*(4), 981–1025. doi: https://doi.org/10.1137/08074489X

Stamm, F. I., Neuhäuser, L., Lemmerich, F., Schaub, M. T., & Strohmaier, M. (2020). Systematic edge uncertainty in attributed social networks and its effects on rankings of minority nodes. arXiv:2010.11546v2 [cs.SI]

Stewart, G. W. (2001). *Matrix algorithms volume 2: eigensystems*. SIAM.

Stewart, I., Golubitsky, M., & Pivato, M. (2003). Symmetry groupoids and patterns of synchrony in coupled cell networks. *SIAM Journal on Applied Dynamical Systems*, *2*(4), 609–646.

Strogatz, S. (2004). *Sync: the emerging science of spontaneous order*. Penguin UK.

Stumpf, M. P., Wiuf, C., & May, R. M. (2005). Subnets of scale-free networks are not scale-free: sampling properties of networks. *Proceedings of the National Academy of Sciences*, *102*(12), 4221–4224.

Tian, F., Gao, B., Cui, Q., Chen, E., & Liu, T.-Y. (2014). Learning deep representations for graph clustering. In *Proceedings of the AAAI Conference on Artificial Intelligence* (vol. 28).

Traag, V. A., Waltman, L., & Van Eck, N. J. (2019). From Louvain to Leiden: guaranteeing well-connected communities. *Scientific Reports*, *9*(1), 1–12.

Trefethen, L. N., & Embree, M. (2005). *Spectra and pseudospectra: the behavior of nonnormal matrices and operators*. Princeton University Press.

Van Lierde, H., Chow, T. W., & Delvenne, J.-C. (2019). Spectral clustering algorithms for the detection of clusters in block-cyclic and block-acyclic graphs. *Journal of Complex Networks*, *7*(1), 1–53.

Von Luxburg, U. (2007). A tutorial on spectral clustering. *Statistics and Computing*, *17*(4), 395–416.

Wagner, C., Singer, P., & Karimi, F. (2017). Sampling from social networks with attributes, pp. 1181–1190. doi: https://doi.org/10.1145/3038912.3052665

Wainwright, M. J. (2019). *High-dimensional statistics: a non-asymptotic viewpoint*, (vol. 48). Cambridge University Press.

Wasserman, S., & Faust, K. (1994). *Social network analysis: methods and applications*, (vol. 8). Cambridge University Press.

Wu, Z., Pan, S., Chen, F. et al. (2020). A comprehensive survey on graph neural networks. *IEEE Transactions on Neural Networks and Learning Systems*. arXiv:1901.00596v4 [cs.LG]

Xu, R., & Wunsch, D. (2008). *Clustering* (vol. 10). John Wiley & Sons.

Young, J.-G., Cantwell, G. T., & Newman, M. E. J. (2020). *Robust Bayesian inference of network structure from unreliable data.* arXiv:2008.03334v2 [cs.SI]

Yu, Y., Wang, T., & Samworth, R. J. (2015). A useful variant of the Davis–Kahan theorem for statisticians. *Biometrika*, *102*(2), 315–323.

Acknowledgements

We would like to thank our many collaborators who have helped us, over the years, to crystallise the ideas presented in this Element. In particular, Mauricio Barahona and Jean-Charles Delvenne have played a pivotal role, as mentors, colleagues, and friends; but we would also like to thank Mariano Beguerisse, Austin Benson, Christian Bick, Vincent Blondel, Alexandre Bovet, Ed Bullmore, Timoteo Carletti, Antoine Delmotte, Tim Evans, Mauro Faccin, Heather Harrington, Julien Hendrickx, Ali Jadbabaie, Juan Kuntz, Sune Lehmann, Cecilia Mascolo, Naoki Masuda, Leto Peel, Mason Porter, Martin Rosvall, Ingo Scholtes, Santiago Segarra, John Tsitsiklis, and Sophia Yaliraki, who have accompanied us in this journey over the years (and we apologise to all those we have forgotten to mention here). We are grateful to the following people for valuable feedback on the manuscript: Bethany Clarke, Karel Devriendt, Florian Frantzen, Maximilian Lutz, Leonie Neuhäuser, and Karl Welzel. Most of this Element was written during the COVID-19 pandemic, a time when lockdowns and school closures transformed our personal and professional lives. We are thus particularly indebted to Bram, Charlie, Thyl, and Charlie for the hours they gave us with relative, intermittent silence in these unusual times.

The Structure and Dynamics of Complex Networks

Guido Caldarelli

Ca' Foscari University of Venice

Guido Caldarelli is Full Professor of Theoretical Physics at Ca' Foscari University of Venice. Guido Caldarelli received his Ph.D. from SISSA, after which he held postdoctoral positions in the Department of Physics and School of Biology, University of Manchester, and the Theory of Condensed Matter Group, University of Cambridge. He also spent some time at the University of Fribourg in Switzerland, at École Normale Supérieure in Paris, and at the University of Barcelona. His main research focus is the study of networks, mostly analysis and modelling, with applications from financial networks to social systems as in the case of disinformation. He is the author of more than 200 journal publications on the subject, and three books, and is the current President of the Complex Systems Society (2018 to 2021).

About the Series

This cutting-edge new series provides authoritative and detailed coverage of the underlying theory of complex networks, specifically their structure and dynamical properties. Each Element within the series will focus upon one of three primary topics: static networks, dynamical networks, and numerical/computing network resources.

Cambridge Elements

The Structure and Dynamics of Complex Networks

Elements in the Series

Reconstructing Networks
Giulio Cimini, Rossana Mastrandrea and Tiziano Squartini

Higher Order Networks: An Introduction to Simplicial Complexes
Ginestra Bianconi

The Shortest Path to Network Geometry: A Practical Guide to Basic Models and Applications
M. Ángeles Serrano and Marián Boguñá

Weak Multiplex Percolation
Gareth J. Baxter, Rui A. da Costa, Sergey N. Dorogovtsev and José F. F. Mendes

Modularity and Dynamics on Complex Networks
Renaud Lambiotte and Michael T. Schaub

A full series listing is available at: www.cambridge.org/SDCN

Printed in the United States
by Baker & Taylor Publisher Services